THE OUTWARD BOUND®
Map & Compass
HANDBOOK

THE OUTWARD BOUND®

Map
&
Compass

HANDBOOK

Revised and Expanded Edition

Glenn Randall

The Lyons Press

Printed in the United States of America

10 9 8 7 6 5 4 3 2

Library of Congress Cataloging-in-Publication Data

Randall, Glenn, 1957–
    The Outward Bound map & compass handbook / Glenn Randall.—Rev. and expanded ed.
        p.   cm.
    Includes index.
    ISBN 1-55821-747-9
    1. Orienteering—Handbooks, manuals, etc.
I. Outward Bound, Inc.   II. Title.
GV200.4.R364   1998
796.58—dc21   98-18603
                                                CIP

Contents

About Outward Bound

Outward Bound is the largest and oldest adventure-based educational institution in the world and is a nonprofit, tax-exempt organization. Best known for its ability to build participants' self-confidence and self-reliance, Outward Bound also develops a sense of compassion for others, responsibility to the community, sensitivity to the environment, and leadership skills in individuals from many different backgrounds.

Outward Bound uses challenging activities, primarily in wilderness settings, to teach both adults and young people more about themselves and others and to help them realize that many of their preconceived limits are self-imposed.

Each Outward Bound course centers around challenging activities for which extensive technical training is given. Included among more than 700 courses offered to people fourteen years of age and older are experiences ranging in diversity from sailing, backpacking, canoeing, sea kayaking, and white-water rafting to mountain climbing, skiing, dogsledding, and even urban experiences. All use the vast majestic wilderness areas of twenty-two states as well as selected urban environments in major U.S. cities. All incorporate traditional Outward Bound elements like map and compass use, orienteering, rock climbing, rappelling, and wilderness camping.

A Brief History

Outward Bound grew out of the need to instill a spiritual tenacity and the will to survive in young British seamen being torpedoed by German U-boats during World War II.

6

From this beginning in 1941, a number of Outward Bound schools were established in the United Kingdom. The movement soon spread to Europe, Africa, Asia, Australia, and North America. Today there are more than fifty Outward Bound schools and centers on five continents.

The first Outward Bound school in the United States was established in 1961 in Colorado. Now there are five U.S. Outward Bound schools: Colorado; Hurricane Island, in Maine; North Carolina; Pacific Crest, in Oregon; and Voyageur, in Minnesota.

Outward Bound has also established a number of urban programs currently operating in Atlanta, Baltimore, Boston, and New York City. These urban programs have been specifically designed to address the needs of inner-city youth and social, cultural, and educational problems existing in every large city throughout the country.

Serving Special Populations

A typical Outward Bound course is from four to nine days for adults, and from three to four weeks for high school and college students. Outward Bound also offers courses specifically developed to serve the unique requirements of some special populations: troubled youth and people with special medical needs, for example. At each school, courses incorporating leadership development and wilderness study may be taken for college credit, and there are also specific courses for professional and managerial groups.

Safety in the Wilderness

All Outward Bound instructors are highly skilled and experienced in wilderness adventure. Training includes the latest technical and safety management practices. Outward Bound's enviable safety record reflects the emphasis each school places on ensuring the well-being of students entrusted to its care.

Benefits

After an Outward Bound experience, participants discover many positive attributes about themselves. All expect more from themselves. They become confident whereas before they were hesitant. They learn to share, to lead and to follow, and to work together as a group. In safeguarding one another, they form bonds of mutual trust. They discover that many problems can be solved only with the cooperation of all members of a group.

Over the past half-century, research has validated these universally recognized positive effects on personal growth. Outward Bound is well known for its ability to enhance confidence and interpersonal relationships, develop leadership skills, and has been shown to provide marked improvements in many other areas of personal and moral development, such as self-esteem, assertiveness, and dependability.

As one Outward Bound student said: "We are better than we know. If we can be made to see it, perhaps for the rest of our lives we will be unwilling to settle for less."

For further information please call, write, or E-mail Outward Bound USA, Route 9D, R2 Box 280, Garrison, NY 10524, toll-free (800) 243-8520, E-mail: national@outward-bound.org. Visit the Outward Bound USA Web site at: http://www.outwardbound.org.

Introduction

IT WAS THE MORNING of our tenth day on Mt. Hunter, and we were in deep trouble. Already our ascent had taken three days longer than we expected and we were still far from home. The mountain's steep ice and steeper rock had sapped our strength, and the frequent storms punctuating the unremitting cold only exacerbated our difficult situation. It was time to get out of there, but the desperate climbing below made it clear we could not descend down the south face. The only way out lay across the summit plateau and down the long and difficult west ridge. The intense cold had frostbitten my fingers and it was threatening to do the same to Peter Metcalf's and Pete Athens's toes. We could have wolfed our remaining food in a single meal, yet we knew it would be many more days before we would eat our fill again. But at least the day had dawned clear, and the first few miles across the summit plateau appeared to be easy.

Shortly after we emerged from our snow cave, our next landmark, the junction of the west ridge and the summit plateau, came into view. We hurried on as fast as the deep snow and our emaciated bodies would permit.

Within an hour, clouds boiled up from below and engulfed us in a soundless white void. Like fools, we had neglected to take a compass bearing on our landmark. But at least we could take a bearing from the map based on our best estimate of our position. I was leading and Pete Athens was in the rear, trying to keep our three-man rope team heading in the right direction by shouting course corrections.

The storm intensified. Athens shouted to me to turn left 90 degrees—a direction that my instinct told me was 90 degrees wrong. For a minute we tried to argue over the screaming wind. Then I gave up and obeyed the compass.

Three hours later, exhausted, with darkness imminent and our position a mystery, we dug another snow cave. As we chipped wearily at the ice, the storm vanished as quickly as it had come. We bolted from the cave. There, only a quarter of a mile away, shining in the low light of the setting sun, was the junction of the west ridge and the plateau. Without a map and compass, the skill to use them, and at least one cool head to insist we should trust them, we might have wandered on the plateau for another day or longer, worsening our frostbite, hastening the deterioration of our already wasted bodies and further reducing the chances of descending safely.

I wrote the first edition of this book several years after my experience on Mt. Hunter, and the lessons learned about wilderness navigation on that climb are just as valid today as they were then. Everyone who plays or works outdoors needs route-finding skills, though only a few, thankfully, will need them as urgently as we did. Day hikers and backpackers will find a map and compass useful for identifying landmarks, making the correct turn at trail junctions and estimating travel time—even if they always follow established trails. Everyone who leaves those well-trodden ribbons of civilization will find route-finding tools essential. When traveling off-trail in dense forest, with the surrounding peaks hidden by trees, a map and compass will keep you on track. Snowshoers and cross-country skiers leave tracks, of course; but in open forests, meadows and above timberline, it takes only a few hours of snow and wind to obliterate those tracks completely. Mountaineers traveling glaciers and featureless snowfields become completely dependent on their route-finding abilities (and foresight in marking their trails) whenever fog and storms move in. Sea-kayakers and canoeists crossing open water must also be accomplished navigators. Even river-runners, who always know in what direction they're heading, need to know how to use a map and compass to find a special side canyon or a particular campsite.

Most of the second edition of this book, like the first, is devoted to teaching the fundamentals of navigating with map, compass and altimeter. The second edition would not have been complete, however, without the addition of a chapter on the Global Positioning System.

The Global Positioning System was originally created by the U.S. military as a way to help its forces navigate. When equipped with a GPS receiver, troops, ships and aircraft can intercept signals beamed to Earth by 24 satellites and know their exact positions anywhere in the world, day or night, regardless of weather. At first, civilian GPS receivers were quite expensive. However, as is commonplace with electronic devices today, the price tumbled and the quality improved. Today's civilian GPS receivers weigh less than a pint of water and cost less than an internal-frame backpack—certainly much less than a half-hour ride on a search-and-rescue helicopter. Equipped with a GPS receiver, you can determine your latitude and longitude to within 100 meters. Take those coordinates to a map and you know where you are. This immensely powerful tool is not without flaws, however; both the benefits and faults are covered in this book.

Learning the craft of backcountry navigation using all the available tools gives you the freedom to roam the wilderness at will, in any season. Navigation skills, particularly the ability to read a map, also give you a ticket to hours of mind travel: poring over possible routes, wondering what secrets some serpentine desert canyon holds, what fish might inhabit some lake buried in the wilderness far from any trail, what elegant climbs some sharp-edged ridge or overpowering granite face may afford. I used to carry a book to read during idle hours, stormbound in camp, but I do so no longer. Instead, I spend my free time poring over my maps, studying the way the contours lie and the streams run, memorizing the names of prominent peaks and lakes and planning my next adventure. With maps and your imagination, the world is your playground.

After reading this book, you'll know how to:

- Create a mental image of a landscape while studying a map;
- Use a compass to find your way to a destination and back again;
- Combine map-reading and compass skills to identify a landmark, plot your course and determine your location;
- Use an altimeter to pinpoint your position;
- Use a GPS receiver to navigate a course from one landmark to the next, then find your way back home again;
- Avoid the most common route-finding errors;
- And use a whole array of easy navigation tricks that simplify staying found in the wilderness.

My father gave me my first compass when I was 12 and I've been practicing getting lost with it ever since. Actually, to paraphrase Daniel Boone, I've never really been lost; I've just been mighty confused. After 28 years of getting confused in far-flung wildernesses from Alaska to Argentina, I've learned route-finding the hard way. If you read this book carefully and (even more important) spend a few hours practicing the skills it teaches, you'll master route-finding the easy way and save yourself a thousand anxious moments in the process.

1

Understanding the Drunken Spider's Web:

HOW TO READ A TOPOGRAPHIC MAP

SOMETIMES YOU CAN SOLVE your navigational problems with just a good map and common sense.

A case in point: a few years ago, a friend of mine and I were sea-kayaking along the storm-wracked coast of Alaska's Kenai Fjords National Park. We had just rounded Cape Aialik, and were midway through a 20-mile stretch of coastline so consistently walled with cliffs that no landing was possible. Cold rain fell intermittently from the swollen clouds only 100 feet overhead. As we headed north from the cape, we stared hard through the fog, eager to identify Porcupine Cove, the first possible landing site.

In the fog, however, all coves looked alike.

"Is that it?" Bill asked an hour after we rounded the cape. He pointed ahead to an apparent indentation in the coast that seemed, in the fog, to have all the credibility of a mirage. I had to agree that the cove looked vaguely right. I'd examined it carefully when we passed it on our way out a week earlier. Then Bill pulled out the map to double-check. "That can't be it," he decided.

How did he reach that conclusion so quickly? In the fog, we could only guess at the cove's outline and dimensions. Although we had been able to see snatches of the coastline as we paddled north, no other landmarks were now visible.

But Bill had landmarks in his head.

"We've only passed two coves so far," he said. "This is the third. Porcupine Cove is still several miles ahead."

That kind of navigating requires no great map-reading skills. It doesn't even demand use of a compass. What it does require is paying careful attention to the landscape around you. Raise your eyes from the ground (or sea) five feet ahead and really study your surroundings! That's the first key to successful navigating.

Being aware means looking back over your shoulder as well as forward. The world looks very different heading down the trail than it did heading up. Memorize the shape of the little pinnacle that marks the point where you joined the ridge. Remember the appearance of the fallen logs that mark where the trail turns away from the sandy desert wash. You can't count on your footprints to guide you when you return along the same trail. A group of hikers could come along, miss the junction and walk half a mile up the wash before realizing their mistake. If you follow the main body of footprints—theirs—you'll make the same error.

Paying attention to your watch can also help you estimate your position. A simple calculation during our attempt to identify the mystery cove would have confirmed Bill's conclusion. We'd only been paddling north from the cape for an hour. We knew from keeping track of hours and mile during the previous week that our average pace was about three miles an hour. Without measuring anything, we could have eyeballed the map and seen that we couldn't possibly have paddled from Cape Aialik to Porcupine Cove in just an hour.

In Kenai Fjords, a basic *planimetric* map would have been adequate. Planimetric maps show the world as if it were all on one plane. A gas station road map is a good example. The map always shows the ground as if the observer was directly overhead. That means a right angle on the ground (a road junction, for example) is always represented by a right angle on the map.

Equally important, one inch on the map always represents a particular distance on the ground. That relationship is the map's *scale*. On a map of the United States, the scale might be one inch to every 200 miles. On a map useful for hiking, the scale might be one inch to every mile or three-eighths of a mile. The scale is usually expressed as a ratio, 1:24,000, for example, which means one inch on the map equals 24,000 inches on the ground. Don't worry, you don't need to figure out what fraction of a mile equals 24,000 inches. As I'll explain in more detail later, most maps have a scale diagram that shows graphically what distance on the map equals one mile on the ground.

Planimetric maps, like most maps, also have a *legend:* a chart showing what the symbols used on the map mean. Some symbols are obvious: irregular blue splotches are typically lakes, for example. Others, like a dashed line, can mean a four-wheel-drive jeep road on one map and a trail for hikers only on another. Knowing the symbol's meaning can prevent you from eating a lot of dust.

Most interesting wilderness areas are not flat. For them, you need a *topographic* map, called a topo for short.

Topographic maps show the ups and downs of the terrain by means of *contour lines*. A contour line is a line on the map that represents the same elevation throughout its length. It may duck into canyons and bulge out around ridges, but it still marks the same elevation. The line indicating the boundary of a lake is a good example of a contour line.

The easiest way to visualize the way contour lines relate to mountains and valleys is to build yourself a little mountain. You'll find the materials in the form of figure 1-1.

First, cut along the outer edge of the figure, as indicated. Place line A atop line B, tape securely, and put the resulting

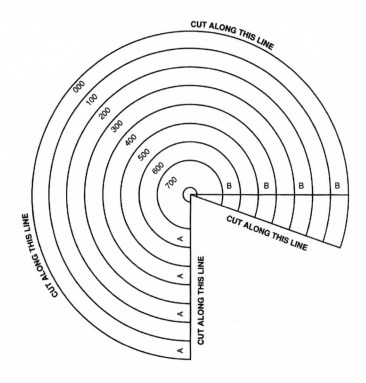

1-1 Cut along the marked lines, place line A over line B and tape securely. Set the cone you've just made point-up on a tabletop. If you'd rather leave the book intact, trace the drawing, then cut out the tracing.

paper cone point-up on a tabletop. Looked at from the side, it should resemble a little volcano with horizontal lines—contour lines—running across it. Each line is at the same height above sea-level throughout its length. The elevation change from one line to the next is called the *contour interval*. On any given map, it's always the same.

Look at the cone from directly above, as if you were in an airplane flying overhead. Notice how the contour lines form concentric circles. The biggest circle is closest to the tabletop. Successively smaller circles represent higher elevations. This view from above shows your paper volcano in exactly the same way that a topo map would depict it. Figure 1-2 shows how your cone would be mapped.

That's the first principle of understanding topo maps: concentric contour lines that form complete, closed paths,

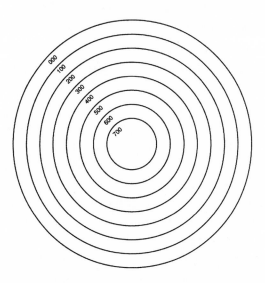

1-2 Representation of your paper cone on a topographic map.

17

whether those paths constitute circles or some irregular shape, represent mountains and hills. Actually, there's one very rare exception. Circular or semi-circular depressions in the earth are represented by similar-looking contours. To distinguish mountains from holes, map-makers draw stubby lines perpendicular to the contours that point to the center of the depression. You can see an example in the list of map symbols in figure 1-3.

To understand how contour lines relate to valleys, look at your paper cone again. Press in slowly on one side of the cone near the bottom to create a valley. Crease the paper so the crease runs along the valley bottom. Now look down at the cone from directly overhead. Let it flatten out in your mind's eye, as if you were looking down at a map. Notice how the contours in the valley form V's whose sharp tips point to higher elevations. Figure 1-4 shows how your volcano-with-valley would be mapped.

That's the second principle of understanding topo maps: contour lines in valleys form V's pointing to higher ground. Sometimes the V's are softened into U's, but the principle still holds.

Ridge contours resemble valley contours except that the contour lines generally form U's and always point to *lower* ground. On figure 1-4, arrows mark the broad ridges that enclose the valley. The presence of a blue line indicating a stream is a sure sign of a valley, but not all valleys have streams. To be certain whether a group of contours represents a valley or a ridge, you need to look for the elevations marked on the map to determine which way the land is sloping.

Look at figure 1-5. The thin contour lines are called *intermediate contours*. The darker, thicker contour lines are

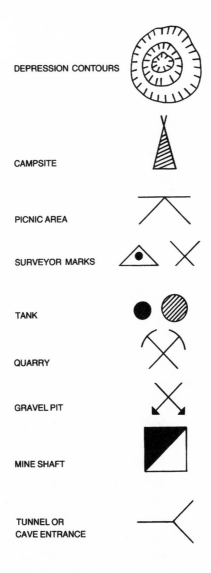

DEPRESSION CONTOURS

CAMPSITE

PICNIC AREA

SURVEYOR MARKS

TANK

QUARRY

GRAVEL PIT

MINE SHAFT

TUNNEL OR
CAVE ENTRANCE

1-3 A selection of topographic map symbols. For a complete listing, see the USGS folder, "Topographic Maps."

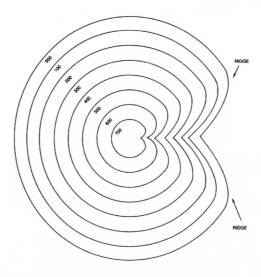

1-4 A topographic-map representation of your paper cone after you crease it to make a valley.

called *index contours*. Every fifth contour is an index contour. Index contours have the elevation they represent written on them at intervals along their length. Elevations always refer to elevations above mean sea-level (the level of the sea averaged over many years and many tidal cycles). To determine which way the ground is sloping at a particular place, locate the two nearest index contours and trace them along until you come to their elevations. Once you know which way the ground is sloping, you can tell whether the bunch of U's or V's you're looking at represent a ridge or a valley.

Very often you'll see the tips of the U's of two ridges pointing to each other. There are two examples in figure 1-5.

20

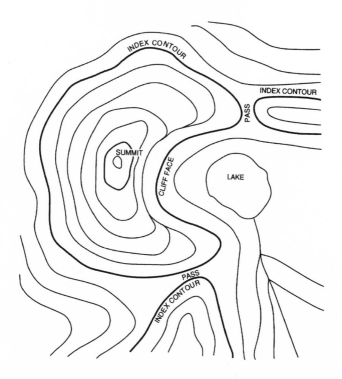

1-5 Thin contour lines are called intermediate contours. Darker contour lines are called index contours. Cliffs and passes are indicated as shown.

This configuration of contour lines represents a pass or saddle. It makes sense: if you started high on either ridge and walked downhill, in the direction the contour U's were pointing, you'd end up at the pass.

The difference in elevation between intermediate contours—the contour interval—is always the same on any particular map. In Florida, the contour interval might be 10 feet; in the Alaska Range, it might be 100 feet. Likewise, the ele-

21

vation difference between index contours is also constant. It's always equal to five times the contour interval, assuming that your map-maker, like most, made every fifth contour line an index contour.

Whether the contour interval is 10 feet or 100, the third topo-map principle still applies: the more closely the contours are spaced, the steeper the terrain. That's because the denser the contours, the greater the elevation change over the same horizontal distance. A vertical cliff is the most extreme example: a large elevation change over zero horizontal distance. Map-makers depict cliffs by drawing contour lines that merge. Figure 1-5 has an example. Another principle: regardless of the terrain's steepness, a path straight up (or down) a slope will always be represented on a map by a straight line perpendicular to the contour lines depicting that slope.

In the United States, the U.S. Geological Survey is the prime source for topographic maps. The USGS publishes many different *series* of maps, each covering a different amount of land. Those most useful to wilderness travelers are the 7½-minute and 15-minute series. No, a 7½-minute map doesn't cover the amount of land you can cross in 7½ minutes, nor does it take 15 minutes to figure out where you are on a 15-minute map. "Minutes," in this case, refers to minutes of latitude and longitude. Sixty minutes makes one degree. A 15-minute map, therefore, always covers one-quarter of a degree of latitude and longitude.

Lines of latitude are lines circling the globe parallel to the equator. They're called parallels. One degree of latitude always equals 69 miles. Lines of longitude, called meridians, pass through the poles and form 90-degree angles with the equator. The distance between lines of longitude varies from about 67 miles at the equator to zero at the poles, where all longitude lines converge. In my home state of Colorado, a 7½-minute map covers an area approximately 8.5 by 6.5 miles.

The area covered by the map—its *series*—does not dictate that any particular *scale* be used. However, the USGS has established some conventions about what scales will be used in each map series. All 7½-minute maps, for example, use either 1:20,000, 1:24,000 or 1:25,000. The 1:20,000 scale is used only in Puerto Rico. All 1:24,000-scale maps give elevations and distances in miles and feet. All 1:25,000-scale maps use metric units. Most 7½-minute maps use 1:24,000, in which one inch equals 2,000 feet or about three-eighths of a mile.

All 15-minute series maps (with the exception of the USGS/Defense Mapping Agency 15-minute quads) use the 1:62,500 scale, in which one inch equals approximately one mile. Alaska is covered by maps at the 1:63,360 scale, in which one inch equals exactly one mile. These maps cover 15 minutes of latitude and 20 to 36 minutes of longitude.

The USGS also makes maps that cover much larger areas, up to and including the entire United States. Confusingly, however, these map series are not designated by the area covered in minutes or degrees, but by the *scale*, such as 1:100,000 or 1:250,000.

While we're on the matter of scale, let's get one other bit of terminology straight: *small-scale* vs. *large-scale*. On a small-scale map, landscape features are shown relatively small. On a 1:250,000-scale map, for example, (which uses a relatively small scale), a mile-long meadow occupies only a quarter of an inch. On a large-scale map, such as a 7½-minute, 1:24,000-scale map, landscape features are shown relatively large. That same mile-long meadow occupies 2⅝ inches. Figure 1-6 summarizes how much territory the different series of maps cover in relation to the state of Colorado. The 1:250,000-series map labeled "Denver," for example, covers about one-sixteenth of the state—an area of approximately 3,500 square miles. That's a map useful for auto travel, not hiking.

It's not important to memorize what one inch means in all these different scales. All USGS maps have a visual reminder at the bottom of the map in the form of a scale diagram showing what distance on the map equals one mile on the land. Before studying a map, check the scale diagram. It will help you visualize the terrain.

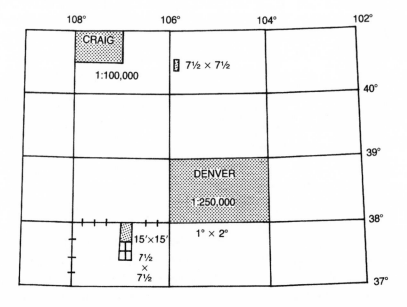

1-6 Topographic maps of different scales and series cover different areas in relation to the state of Colorado.

Fortunately, the USGS is a lot more consistent in its use of colors and symbols than it is in its use of scales in a particular series of maps. Black, for example, always indicates a man-made feature such as a road or building. Blue, logically, denotes water: streams, rivers, lakes, the ocean. Brown is used for contour lines. Green indicates vegetation: brushy and forested areas. Red designates major highways. It's also used for the boundaries of units of public land: national park boundaries, for example, or townships and sections of land. Purple indicates features added in the map on the basis of serial photographs. Purple features have not been field-checked, which means nobody visited the site in person and verified what the aerial photograph appeared to show.

The USGS uses several dozen symbols on its maps to indicate everything from surveyor's bench marks to standard-gauge multiple-track railroads. Figure 1-3 shows some symbols. Each USGS map shows the road classification symbols. For a complete listing, get the USGS's folder called "Topographic Maps." It's free. I'll give you the address when I tell you how to order maps.

Topo maps depict unchanging physical features very accurately. But as time passes since the date of printing, some natural features are likely to evolve. A lake, for example, may become a meadow; a glacier may advance or recede. Man-made features are also likely to change. Roads may be upgraded or abandoned. New trails may be added, or old ones re-routed. If you're visiting a national park or national forest, you may be able to keep up with that kind of change by obtaining an up-to-date planimetric map from the park or national forest administration. A friendly ranger might also help you update your topo map.

You can buy topo maps directly from the USGS as well as from many outdoor specialty shops. To determine the ones you need, obtain an Index to Topographic Maps for the state you're interested in. It's free if you order it from the USGS. Most retail stores that stock maps have a shop copy.

The index is a map of the state that shows every topo map published that covers any part of the state. (Some topos cover more than one state.) Each rectangle on the index represents a topo map. The name of the map and the date of the survey are written within the rectangle. In general, the smallest rectangles on the index will be 7½-minute maps; in Alaska, they'll be 1:63,360 maps showing 15 minutes of latitude. You can check by counting the number of rectangles between parallels of latitude. If there are eight per degree, the rectangles show 7½-minute maps because eight times 7½ equals 60 minutes or one degree. Four rectangles per degree means the maps show 15 minutes of latitude.

Maps are named for some prominent feature within their boundaries. Often that will help you decide if you need that map. Indexes also show major roads, towns, rivers and mountains, which may help you identify which maps you need. When in doubt about a map, order it anyway. Even if your intended route doesn't cross it, you'll probably be looking out at those peaks and valleys when you stop for lunch on the summit. Be sure to specify the state, the quadrangle name and the map series when ordering. In some cases, a 15-minute map may have the same name as a 7½-minute map contained inside it, which makes the series designation crucial in obtaining the right map.

In Alaska, only the 1:250,000 maps have names. The 1:63,360 maps have only letter and number designations. To order a 1:63,360 map, give the name of the 1:250,000 map that contains it, plus the letter and number. An example might be Fairbanks D-1. If you have a map that covers part of the territory you're interested in, look at its margins. They'll tell you the name of all eight adjoining maps, including those that only touch at one corner.

If you want maps of areas east of the Mississippi (that includes Minnesota, Puerto Rico and the U.S. Virgin Islands) order from:

Eastern Distribution Branch

U.S. Geological Survey

1200 S. Eads St.

Arlington VA 22202

If you want maps of areas west of the Mississippi (that includes Alaska, Hawaii, Louisiana, American Samoa and Guam) order from:

Western Distribution Branch

U.S. Geological Survey

Box 25286, Federal Center

Denver CO 80225

Alaska maps only are available from:

Alaska Distribution Section

U.S. Geological Survey

Box 12, Federal Building

101 Twelfth Ave.

Fairbanks AK 99701

Order Canadian maps from the Canada Map Office, 615 Booth St., Ottawa Canada K1A 0E9. Ask for a free Index of National Topographic System Maps for your area of interest. Index 1 covers the Maritime Provinces, Quebec and Ontario; index 2 covers Manitoba and the western provinces; index 3 covers everything north of the 60th parallel.

Mail-ordering anything is a pain, and maps are no exception. Each state index lists all map dealers in the state. With a little luck, one will be near you. Then you can go and examine maps before buying to be sure you get the right one. Most map dealers carry compasses as well. Many also carry altimeters.

State indexes also list all the special maps that cover part of the state. In Colorado, for example, Rocky Mountain National Park, Mesa Verde National Park and the Black Canyon of the Gunnison, among other interesting places, all rank special maps of their own. Ordering one of them instead of four or five individual sheets can save you a lot of paperwork when you're out in the field. The disadvantage may be that the scale is smaller.

Maps can tell you two things you really want to know before you launch yourself into the wilderness: Can I get there from here? And, how long will it take?

If a trail leads to your destination, then all you need to know is the time required. First, calculate the horizontal distance. A ruler makes it easy to measure the straight-line distance, but trails never run straight. In fact, they're notorious for taking a lot more miles to reach a destination than the straight-line distance would indicate. To calculate distance along a trail, use a bit of string or a pipe cleaner. Place one end of the string on the trailhead, then trace the trail with the string, following the trail as it twists and turns. Use your thumbnail to mark the point where the string crosses your destination, then bring the string down to the scale at the bottom of the map and read off the distance. It's a good idea to add a pessimism factor of 10 percent to account for the curves you shortcut and the switchbacks the map maker neglected to show.

As an alternative measuring tool, use the edge of your compass. One edge on many models is marked in inches. Use short, straight distances to approximate curves. As a third alternative, buy a map-measuring device. The most common ones have a wheel which you roll along the map surface. A pointer indicates miles or kilometers for several different map scales.

Measuring distances by any method is easiest on a flat surface, like your kitchen table during the evening before a day hike, or your tent floor before you strike camp.

After determining the mileage, calculate the elevation change. First, determine your starting elevation. Find the index contour closest to the trail head, yet still below it, and trace it along until you find a place where the map gives its elevation. Then count the number of intermediate contours between your index contour and the trail head. Multiply the number of intermediate contours by the contour interval (it's given at the bottom of the map) and add that number to the elevation of the index contour. For example, let's say the closest index contour below the trailhead is at 10,000 feet. The contour interval is 40 feet, and the trailhead is three contours above the 10,000-foot index contour. You're at 10,120 feet.

Determine the elevation of your destination the same way. Subtract, and you have your elevation gain or loss.

In some cases, you may find that the map doesn't give an elevation for the first index contour below the trail head. In that case, find the nearest index contour that does have an elevation indicated and calculate the elevation of the index contour near the trailhead. If every fifth contour is an index contour, as is normal, then the elevation difference between index contours equals five times the contour interval. For example, if the contour interval is 40 feet and every fifth contour is an index contour, then the elevation change between index contours is 200 feet.

If the trail runs up and down, be sure to add up all the individual segments of elevation gain.

Now you have the most important information for estimating travel time: distance and elevation gain. But there are other variables as well: your fitness, your load, the altitude, the roughness of the trail, the depth of snow (if any), your mode of transportation (foot, skis or snowshoes) and whether you're out for the exercise or to admire wildlife. Two miles an hour is a reasonable pace on a level trail with a moderate load. For every 1,000 feet of elevation gain, add an hour to the time you calculated for the mileage alone. That'll give you a very rough estimate. Then go out and hike and ski some trails! Write down your times over different kinds of terrain. Make sure you add in rest stops. You'll soon learn what's reasonable for you.

If no trail leads to your destination, getting there by any route may be problematical. Your map can give you important clues to the easiest route, but only an actual visit—or a call to a knowledgeable local—will tell you if your planned route is feasible. Those green areas, for example, may be open woods allowing pleasant strolling; they may also be nearly impenetrable thickets of alder, willow and devil's club. The most important information the map gives to those planning an off-trail hike is probably the average angle of the steepest part of the route.

Find your route's steepest part by locating the place where the contours are closest together. Then measure that horizontal distance. USGS maps have scales in both miles and feet. In this case, use the feet scale because you'll be determining the elevation in feet.

Next, measure the vertical rise over that same horizontal distance.

If the vertical rise is the same as the horizontal distance, you're looking at a 45-degree slope that's likely to be very tough walking. In fact, it probably involves stiff scrambling, at least in places, and it may confront you with some real cliffs. If the vertical rise is greater than the horizontal distance, you're most likely confronting mountaineering territory where you'd be advised to have a rope, some hardware and some well-honed climbing skills.

When the rise is only half the horizontal distance, you're looking at a slope of 27 degrees—steep hiking, but only hiking. You can probably find your way around any clifflets that may be hiding in between the contour lines.

Like all generalizations, these have exceptions. In the canyon country of the American Southwest, for example, certain geological strata routinely form cliffs. The average angle of the slope climbing out of a canyon may be quite moderate, but the steepest angle is often vertical. Short but impassable cliffs can run for miles without a navigable break. The same is true in parts of the Canadian Rockies and Colorado's San Juans. Once again, local experience is the real key.

How long will a cross-country route take you? If you're bushwhacking or scrambling amidst cliff bands, the time required is anyone's guess. It could take two hours or more to travel a mile. On the other hand, if you're strolling along a smooth, level ridge above timberline, a mile might take you only half an hour. Traveling on skis off-trail is even slower than traveling off-trail in the summer because of the extra effort of making a track through the snow. In the summer, if you're fit, carrying only a daypack and moving hard, you may be able to gain 1,000 feet in an hour, at least for a while. In most cases, you're likely to do less. If your route combines trail walking and cross-country travel, you'll usually save time if you stick to the trail until you're as close as possible to your destination, then head straight for it. Leaving the trail when you first catch sight of the peak you want to climb and heading diagonally up a steep slope usually takes longer than walking the trail until you're directly below the summit. As with trail walking, you have to spend some time in the country to learn how map features translate into terrain and what your personal capabilities are.

2

The Magic Needle:

CHOOSING AND USING A COMPASS

A VICIOUS SQUALL was raking the treeless tundra below Alaska's Mt. Sanford. Snow driven by a 40-mph gale blew horizontally past the cabin windows. The quarter-inch-thick guy wires supporting the cabin whistled and shrieked as the wind dealt body punches to the flimsy plywood walls. Visibility—between the gusts—sometimes reached 50 yards. Somewhere out in the maelstrom, Chris Haaland and Sara Ballantyne were carrying a load of mountaineering gear toward the foot of Sanford's Sheep Glacier.

Only rolling brown hills dissected by wandering streams lay between the cabin and the glacier. All looked identical, particularly when viewed through ice-encrusted lashes and a punishing veil of flying snow. To make the route-finding even more difficult, Chris and Sara planned to return to the cabin at night, after caching their equipment.

But Chris and Sara had one thing going for them. They knew approximately where the cabin was on the map and where they wanted to go. Based on that knowledge, they had taken a *course* off the map. In other words, they had used a compass to measure the *angle* between the true north and the direction they wanted to go.

As soon as they stepped from the cabin, they sighted along their course as far as they could see, picked a landmark and headed in that direction. Whenever they reached a high point—a ridge or hilltop—they built a small pile of stones, called a cairn, to aid their return.* Then they sighted along their course again, picked another landmark, and continued.

Eventually they felt rather than saw that they had reached a low pass. Below them lay Sheep Glacier. They cached their loads, turned around, and began following their cairns towards home. Without the cairns to correct their course periodically in the rough terrain, they might well have wandered away from their route without realizing it. At dusk, with the storm still pounding the tundra outside, they walked into the welcome shelter of the cabin.

Chris and Sara had navigated across six and a half miles of tundra—and back—using nothing but a compass, a few piles of rocks and an initial course taken off the map. Once the storm closed in, the map was of little further use. If they had had even a few minutes' visibility before the squalls blew in—just long enough to take a compass bearing off their destination—they wouldn't even have needed the map. Ingenuity and a compass are often enough by themselves to find your way.

The most basic compass is simply a magnetic needle suspended on a pivot so the needle can align itself with the earth's magnetic field. The needle is usually mounted in some kind of circular housing marked with the cardinal directions: north, south, east and west. Although even the simplest compass lets you roughly determine directions, for precise navigation you need something more sophisticated.

*Cairns should be used only when absolutely necessary, and should be knocked down and scattered during your return trip so that no evidence remains of your passage.

First, you need a compass that lets you sight some landmark in the field and take its *bearing.* A bearing is nothing more than the angle between a line heading north from your position and a line heading towards the landmark. (The north line can be defined as true north, the direction of the North Pole, or magnetic north, the direction the compass needle points. I'll get into that difference later.) For now, study figure 2-1 and memorize this concept: a bearing is an *angle,* measured in degrees, between a line heading north and a line heading toward your landmark. You always start counting at the north line, and you always count *clockwise* around the compass dial.

Second, you need a compass that lets you measure angles on a map. Once again, the angle of interest will be between a

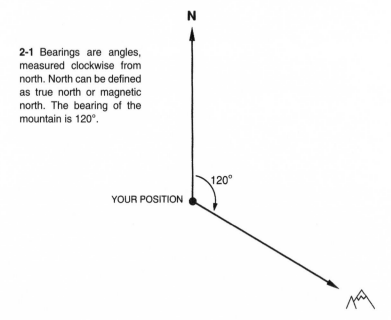

2-1 Bearings are angles, measured clockwise from north. North can be defined as true north or magnetic north. The bearing of the mountain is 120°.

2-1 cont. The bearing of the tree is 40°.

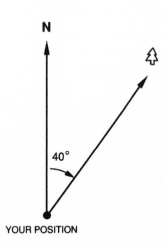

2-1 cont. The bearing of the lake is 320°—not 40°. (Always measure clockwise around the scale from north.)

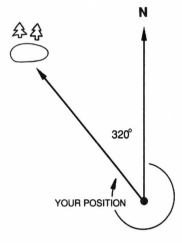

line running north, and a line running through your position and some mapped landmark. I'll call that angle a *course*, since you will normally use that angle to pick the course you will travel. For that purpose, the compass needle is unnecessary. What you really want is a *protractor:* a device for measuring angles.

Since it's awkward to carry around both a compass and a protractor, compass manufacturers have combined the two instruments in one. They're called *protractor compasses,* or, more commonly, *baseplate compasses* and they're fundamental to most of the navigational techniques discussed in this book. Figure 2-2 shows their basic parts.

The baseplate compass is named for its clear rectangular base. The circular housing for the compass needle is mount-

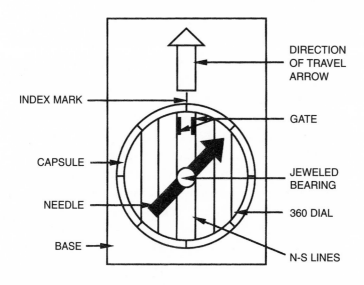

2-2 Parts of a baseplate compass.

ed at one end of this rectangular base. Let's call that circular housing the *capsule*. The capsule rotates in relation to the baseplate. The outer edge of the capsule is marked north, south, east and west and is also marked in degrees, increasing as you move around the capsule clockwise. Zero degrees equals north; 90 degrees equals east; 180 degrees equals south; 270 degrees equals west and 360 degrees once again equals north, or zero. Inside the capsule, you'll usually see a series of parallel lines. They're called north-south lines because they lie parallel to a line running through the north and south points on the capsule.

Most baseplate compasses have an arrow inscribed on the base. It's called the *direction-of-travel* arrow because it indicates the direction you want to go when the capsule is set to a course you've taken off the map. You point the same arrow at a landmark in the field when you want to measure a bearing. The foot of the direction-of-travel arrow serves as the *index mark* where you read off the number of degrees to which the capsule is set. To set the capsule to any angle between zero and 360 degrees, turn the capsule until the correct number of degrees is aligned with the index mark.

The compass needle rotates within the capsule, coming to rest when the north end of the needle points to the magnetic north. Many map and compass operations require you to twist the capsule until the north end of the needle points to the north mark on the capsule. To make it easy to determine when the needle and the capsule's north mark are accurately aligned, the capsule has a long, thin rectangle or box inscribed on its bottom surface. I'll call that rectangle the *gate*. Later in this book, when I say, "place the needle in the gate," I mean to align the needle so its north end points to the north mark on the capsule.

Together, those three moving parts—baseplate, capsule and needle—let you measure the bearing of a landmark in the field and measure courses on the map. I'll tell you how after I finish discussing compasses.

Some baseplate compasses have a mirror attached to the baseplate with a hinge. When you're taking a bearing, that mirror lets you see both the object you're sighting and the capsule at the same time. That increases the accuracy of your bearing measurement. The mirror also lets you admire your handsomely weatherbeaten features after a week in the backcountry—greasy hair, burnt cheeks and all. For the first reason only, I recommend a mirror model.

As I've already mentioned, compass needles almost never point to true north. The difference in direction between true north and magnetic north is an angle called the *declination*. All USGS maps state in the bottom margin the declination for the region covered by the map. Compensating for the difference between magnetic and true north when doing map and compass work with a standard baseplate compass requires some simple addition or subtraction. When you take a bearing in the field, you've measured an angle between *magnetic* north, indicated by the compass needle, and your landmark. When you take a course off a map, however, you've measured an angle between *true* north, indicated by the top of the map, and your landmark. You have to reconcile the two means of measurement, a process I'll describe later.

If you'd rather not mess with that, you can buy a compass that lets you set the declination and forget it. On standard baseplate compasses, the gate marked on the capsule is fixed in relation to the capsule. When you place the north end of the needle in the north part of the gate, the needle points exactly at the north mark on the capsule.

"Set-and-forget" compasses, on the other hand, have a gate that moves in relation to the north mark on the capsule. You set that moveable gate at an angle to true north equal to the declination and forget about it until you travel to some different region, where the declination is different. Now when you place the needle in the gate, the needle points to magnetic north and the north mark on the capsule points to true north. See figure 2-3. There's no need to do any addition or subtraction, or even to remember what the declination equals. Set-and-forget compasses are indeed easier to use. I recommend them.

Every compass worth buying will be fluid-dampened. That means a fluid within the capsule prevents the needle

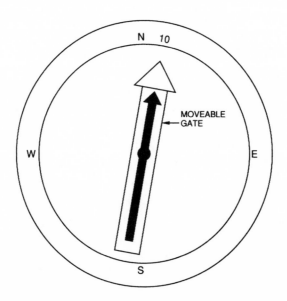

2-3 A set-and-forget compass set for a declination of 10 degrees east. The needle points to magnetic north; the north mark on the capsule indicates true north.

from swinging to-and-fro like a hypnotist's watch. Good compasses will still function at 40° below zero F, although it's best to keep the compass inside your coat to prevent bubbles from forming as the liquid contracts. High altitude, particularly when combined with severe cold, can also cause a bubble to appear. The bubble won't interfere with accuracy unless it's bigger than a quarter-inch in diameter. Usually the bubble will disappear when you return to normal temperatures and sea-level. If it doesn't, you may have a leak. Avoid buying a compass that already has a bubble in the capsule.

You should also protect your compass from extreme heat, such as the dashboard of your car. High temperatures cause the liquid to expand, which can rupture the capsule.

Be sure to buy your compass in the region where you intend to use it. Only along a line known as the magnetic equator, which lies near the geographical equator, do compass needles sit level with the earth's surface. North of the magnetic equator, they dip to the north; south of the magnetic equator, they dip to the south. Manufacturers compensate by counter-balancing the compass needle so that the needle pivots freely when the compass is held horizontally. Silva, for example, makes compasses balanced for five different magnetic zones. If you take a compass balanced correctly for the United States to Australia, the needle is likely to bind.

If you buy a mirror-sight compass, be sure the design lets you see the capsule while you sight landmarks both above and below your elevation. You can check in the store just by sighting something on the floor a few feet in front of you, then sighting something on the ceiling, and making sure you can see the capsule and needle while you're sighting.

You can buy compasses with any number of different scales marked along their edges. While those scales can be handy if they happen to match the scale of the map you're using, they're not a sufficient reason to buy a particular compass. After all, you can always photocopy the scale off the map you'll be using on your next trip and tape it to your compass.

It'll make my job—and yours—a lot easier if you go out and buy yourself a baseplate compass and a topo map before you continue reading this book. There's nothing like having the real thing in your hands to make all this talk of aligning needles in gates and twisting capsules to different bearings seem as simple as it really is. If a picture is worth a thousand words, then actual experience is worth ten thousand.

Now that you've got a map and compass in your hands, you're about ready to use it. First, though, you need to know about a common pitfall. Dale Atkins, a mission leader for Colorado's Alpine Rescue Group, tells a story that illustrates the point perfectly.

One day several years ago a group of hunters drove to the trailhead below Mt. Evans, got out of their truck and spread out their map on the hood. They carefully oriented the map with the compass and just as carefully followed the direction the compass needle said was north. That night, when two teenage members of the group did not return, the adults reported them missing. Dale and a search team from Alpine Rescue drove to the roadhead.

"They went north," the adults said, pointing toward a small peak. Dale walked 30 feet away from the truck and got out his compass.

"But north is that way," he said, pointing in a direction 45 degrees away from the peak. Fortunately, Dale's search team found the teenagers unharmed.

The hunters' mistake? Using their compass on the hood of their truck. Any metal object, even much smaller ones like pocket knives and belt buckles, can throw off the compass needle. The same is true of all electronic devices. Any time current flows through a wire, it generates a magnetic field. That field can affect the compass needle.

Before using your compass, therefore, make sure nothing metallic or electronic is affecting it. Once you've done that, trust it. Only once in all my years of compass-use have I set my compass down on a rock and found that some metallic ore in the rock was affecting the needle. Simply standing up and walking a few feet freed the needle from the rock's influence.

The first technique to learn is how to take a bearing off a landmark, then walk that bearing. Since you're not, at this time, relating your bearings to a map, there's no need to worry about the difference between true north and magnetic north. You can simply relate all directions to magnetic north.

Let's assume you bought a baseplate compass without a mirror sight. To take a bearing off a landmark in the field, point the direction-of-travel arrow at the landmark. Hold the compass level, so the needle can swing freely. Now hold the baseplate still and rotate the capsule until you've put the needle in the gate—in other words, until the north end of the needle points to the north mark on the capsule. Then read the bearing at the index mark. Figure 2-4 shows you how. In this example, the bearing is 316 degrees.

All that you've done is to measure the angle between a line heading toward magnetic north, as indicated by the compass needle, and a line heading toward your landmark, as indicated by the direction-of-travel arrow. The angle is always measured clockwise around the compass dial, starting at north.

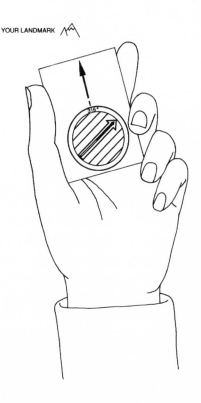

2-4 To take a bearing on a landmark, point the direction-of-travel arrow at the landmark and place the north end of the needle in the north end of the gate. Read the bearing at the index mark.

Taking a bearing with a mirror-sight compass is essentially the same. The only difference is that the mirror-sight compass allows you to sight on your landmark more accurately.

You can use a mirror-sight compass just like a regular baseplate compass. To do so, open the mirror housing all the way, until the compass lies flat. On most models, a line will be inscribed on the mirror. It lies parallel to the long edges of the compass and starts at a notch in the center of the cover's top edge. That line serves as the direction-of-travel arrow. Then proceed as you would with a standard compass.

For more accuracy, however, hold the compass level at about the height of your chin. You'll have to adjust that height a bit if you're sighting an object either above or below your position. Now open the mirror cover until you can see the capsule in the mirror. Hold the compass so the line inscribed on the mirror passes directly through the needle's pivot point as you look at the reflection of the capsule. Doing that ensures that the long edges of the compass point directly at the landmark. Now sight the landmark through the notch in the center of the cover's top edge. Place the needle in the gate by twisting the capsule. Double-check that the line on the mirror is still passing through the needle's pivot point. Read the bearing at the index mark.

Now you've got a bearing on your landmark. Let's say that you, like Chris and Sara, want to walk toward this particular landmark. Just after you start, a storm rolls in, so you're forced to navigate by compass. Or perhaps you've climbed to a mountain-top and want to take a different way down. You can see your destination—a lake, let's say—but shortly after you drop off the summit, you'll be in thick woods and the lake will be invisible. You could just start out walking the bearing, figuring you'll get there in an hour or two. But walking in a straight line in rough terrain is a lot tougher than you'd expect. Once you've got a bearing on your distant landmark, sight through the compass again, but this time, pick a landmark close at hand. In really foul weather or in the woods, that might mean something only a hundred yards away. Walk to that object, sight again, and pick another landmark at the limit of good visibility. Continue until you reach your destination.

Let's say you encounter some obstacle you can't walk over as you follow your compass bearing. It could be a lake; it could be a stream where you'll have to walk up or down to find a ford or fallen log; it could be a wide crevasse in a glacier or a hill ringed with cliffs. If you can see across the obstacle, the solution is simple. Sight along your bearing to some object on the far side of the obstacle, do whatever you have to do to get around the obstacle, and walk to the object you sighted. Pick another landmark along your former line of travel, and continue.

If you can't see across the obstacle—the fog's too thick, let's say, or you're confronting a cliff—then you'll have to maintain your sense of direction while you skirt the obstacle. First, decide if you want to go around the obstacle to the right or left. Let's say right looks easier. Make a right-angle (90-degree) turn to your right and walk far enough to clear the obstacle. Count your steps as you go. Now turn 90 degrees to your left, which puts you back on your original bearing, and walk until you're sure you've cleared the obstacle. There's no need to count steps here. Make another left-hand 90-degree turn and walk the same number of steps you counted after your first turn. That puts you back on your original line of travel. Turn right, sight a landmark on your original bearing, and proceed. All you've done is walk three sides of a rectangle so that when you arrive on the far side of the obstacle, you're back on your original line of travel. Figure 2-5 makes all this clear.

To simplify making all those 90-degree turns, use the short edges of the baseplate to sight landmarks. To make the initial 90-degree right turn, for example, face in roughly the correct direction and sight along a short edge of the baseplate while keeping the needle in the gate. Pick a landmark and go,

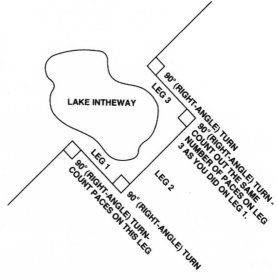

2-5 Maintaining your course while circumventing an obstacle.

counting your steps. When you've cleared the obstacle, make the first left turn by sighting an object along your original bearing, using the compass in the normal way. To make the second left turn, face in roughly the correct direction and sight again along a short edge of the baseplate with the needle in the gate. Pick a landmark, pace out your measured distance and you're back on track.

To return home, you want to walk in exactly the opposite direction—180 degrees opposite. To do so, you can either add or subtract 180 degrees to your original bearing. Use whatever operation keeps the result between 0 and 360. Or you can align the south end of the needle in the north end of the gate on the capsule. Either operation will point you toward home. You then pick landmarks and walk toward them in exactly the same way you did heading out.

3

Putting the Needle to the Drunken Spider:

USING COMPASS AND MAP TOGETHER

THE CALL WAS URGENT: two mountaineers were overdue on James Peak. Given the vicious midwinter storm pummeling Colorado's Front Range, overdue probably meant lost and in trouble. The dispatcher for Alpine Rescue, who had received the call, immediately paged all available members.

Dale Atkins met several teammates at the trailhead and formed a search plan. They would begin by climbing St. Marys Glacier. Then they would head west, across the treeless alpine barrens toward the 13,294-foot summit. When they finished searching there, they would turn to the north and descend to James Peak Lake. Snowmobiles would meet them at the lake and take them back to the road.

A westerly gale made every step a struggle as they started up St. Marys Glacier. The unrelenting wind kicked up a ground blizzard that slashed visibility to a few feet. The worst gusts struck like tidal waves, threatening to knock them flat. All the searchers possessed tremendous winter mountaineering experience, however, and were intimately familiar with the James Peak area. There seemed to be no need to fool around with their maps and compasses, which remained securely stowed inside their packs. They completed their search of the region above timberline without locating the victims, turned to the north, and descended to the lake.

No snowmobiles! Surprised, they radioed their dispatcher. The snowmobiles, they were told, were at the lake and waiting. "Where are you guys?" the dispatcher asked.

The search group started walking around the lake. A dam emerged through the flying snow. Suddenly a sickening feeling of recognition flashed into their minds. Somehow they had turned south near the summit, not north, and descended to Loch Lomond, on exactly the wrong side of the mountain. Sheepishly, they walked back to their vehicles. The lost mountaineers hiked out a day later, uninjured. Every winter thereafter, Alpine Rescue's members gave the mission leader for that search a roll of white surveyor's tape so he could mark his trail and find his way home.*

As those hapless searchers demonstrated, the best map and compass skills in the world are worthless if you don't pull out the tools and use them. It can be really hard to locate your position if the first time you check your map is several hours after you wander away from the last landmark you can identify. It's particularly hard to find yourself if you don't keep track of the direction you're traveling. That sounds obvious, but as Dale and company showed, even the best can get cocky about their ability to navigate without navigational tools.

The best place for your first map check is the trailhead. First, find your location on the map. Usually that's easy:

---

*Although surveyor's tape can be a valuable tool in some search and rescue operations, it should always be removed when the search is concluded to prevent degradation of the land's wilderness character. It should not be used in ordinary routefinding activities.

you're at the end of the road, or at the marked point where the trail leaves the highway. Sometimes, though, it's not so obvious, such as when you're starting a cross-country hike. If you haven't parked near some obvious, mapped landmark, you may need to use the more sophisticated techniques I'll describe later to pinpoint your position.

Next, orient the map. In other words, place the map on some flat surface so that directions on the map correspond to directions in the field. It's always easiest to visualize what you're doing if the map is oriented. Sometimes you can orient the map by eye. If there's a lake straight ahead of you, and a prominent hill to your right, twist the map until the lake lies straight ahead of your position on the map and the hill is to the right of that position. Here's another way to look at it: a map is oriented correctly if a line drawn on the map from your map position to the mapped lake points straight toward the real lake. In similar fashion, all other directions on the map will also correspond to reality.

Sometimes, though, you can't orient by eye. The landscape may lack any obvious mapped features. Or it may have too many features, and you can't tell which is which. In that case, you need to use your compass to orient the map.

Here, and for the rest of this chapter, I'm going to assume that magnetic north and true north lie in exactly the same direction. That assumption will make it a lot easier for you to learn. In the next chapter I'll tell you how to compensate for the difference that usually exists between those directions.

On almost all maps, north is at the top of the sheet. To orient the map roughly, hold the compass horizontal and glance at the needle. It points north, of course; twist the map so the top also points north. Note that when the map is oriented correctly, the left and right margins will represent lines running north and south.

To orient the map more accurately, set the capsule to zero degrees. Place the compass on the map so one long edge of the baseplate lies atop either the left or right margin. Now rotate map and compass together as a unit until you've placed the needle in the gate, so the north end of the needle points to the north mark on the capsule. The map is now oriented. Every direction on the map corresponds to directions in reality. Figure 3-1 shows a correctly oriented map.

Now that you've found your position and oriented the map, take a look around and identify some nearby landmarks. Determine your general direction of travel. Does your route run north, south, east or west? Try to develop a feel for the relationship between the cardinal directions and major terrain features. You might note, for example, that the valley you'll be hiking up runs east and west, while the region's highest peak is basically to the north. Knowing where you started and what direction you're traveling will help prevent dumbness attacks like plac-

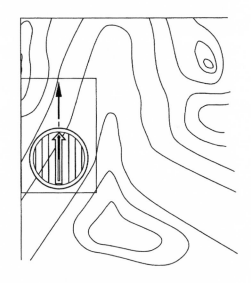

3-1 Orienting the map.

ing the south end of the compass needle in the north end of the gate.

And while you've got the map out, try to create a mental image of the terrain you'll be traveling through. Maybe you head east up that creek for two miles, then make a sharp left just past a big cliff, leaving the main trail and following a spur trail that climbs toward a pass to the north. Knowing that kind of thing will help you start looking for a trail junction at the appropriate time. Don't count on a sign to jolt you into looking up from your companion's boot heels.

Make it a habit during your hike to get out your map and compass every hour or so and locate your position on the map. Keep track of the time you started and the time it takes you to reach various landmarks—that trail junction, for example. It'll give you a sense of your pace that day, which will help you keep track of your location.

Let's say your goal for the day is the summit of some peak. You reach it at lunchtime and sit down to admire the enormous panorama of mountains spread out before you. Orienting your map will give you a rough idea which peak is which. If you want to know accurately, however, you'll need to get a bit more sophisticated.

First, take a bearing on the peak you're interested in using the technique I described in the previous chapter. A glance back at figure 2-4 on page 44 should refresh your memory. That bearing, you'll recall, is just the angle between a line heading north and a line leading to the peak. The angle is measured clockwise from the north line. Although you used the direction-of-travel arrow or notch in the mirror to sight the object, note that you would have gotten exactly the same result if you'd sighted along one long edge of the baseplate.

Now you're going to transfer that angle to the map. One line of the angle, represented by the compass gate, will point north; the other line, represented by one long edge of the baseplate, will run right through your position and toward the peak you're interested in.

To make the logic of the next step easier, orient the map before continuing, then reset the compass to the bearing you just measured. Now place the compass on the map so that the compass gate points north, toward the top of the map, and so that one long edge of the baseplate sits on top of your position. Note that the long edges of the gate (and the north/south lines in the capsule) run parallel to the right and left margins of the map. Don't twist the capsule in relation to the baseplate. Ignore the compass needle. You're simply using the compass as a protractor now, so the needle is irrelevant.

The long edge of the baseplate that is sitting atop your position now points directly at the peak you're interested in. Figure 3-2 shows a compass placed correctly on the map. Note—this is important—that you must make sure you follow the long edge of the baseplate in the direction indicated by the direction-of-travel arrow. With the map oriented correctly, you'll see that the long edge of the baseplate not only points to the paper mountain—it also points to the real one.

Often the edge of the baseplate isn't long enough to reach from your location to the peak of interest. You need an extension: a plastic ruler. Not a metal one; you don't want any metal sirens singing to your needle. Place one edge of the ruler alongside the edge of the baseplate that runs through your position. The ruler will now extend out towards the peak of interest, making it easy to identify the correct one.

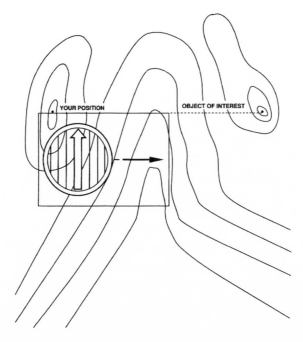

3-2 Applying a bearing taken in the field to the map. Make sure that one long edge of the compass sits on your position, that the north/south lines run north and south and that the north end of the capsule points to the north end of the map.

Your map and compass will also help you solve the opposite puzzle. Once again, let's assume you know where you are on the map. This time, though, you want to use the map to tell you which way to go. We faced this situation on the summit plateau of Mt. Hunter, for example. You'd face a similar problem if you were down in the woods at Lake Hereweare and wanted to know the direction to Lake Overthere.

To solve this problem you need to measure an angle on the map and transfer it to the terrain. The angle, called a course, will be the angle between a line heading north and a

54

line heading to your destination, with your position as the point of the angle.

Start by laying one of the long edges of the baseplate along an imaginary line connecting your position and your destination. Sometimes, as in figure 3-3, the baseplate will be long enough to extend between the two; other times you'll need to get out your ruler again to position the baseplate accurately. Be sure the direction-of-travel arrow points at your destination. Now twist the capsule until the gate points north, as shown in figure 3-4. Both the long edges of the gate, and the north/south lines, will run parallel to the left and

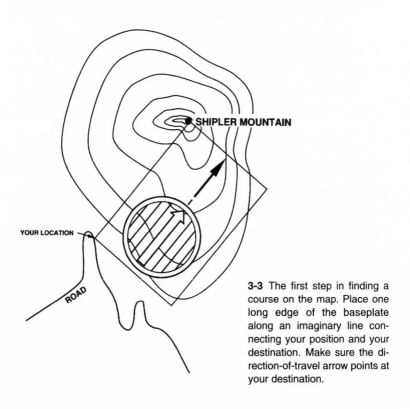

3-3 The first step in finding a course on the map. Place one long edge of the baseplate along an imaginary line connecting your position and your destination. Make sure the direction-of-travel arrow points at your destination.

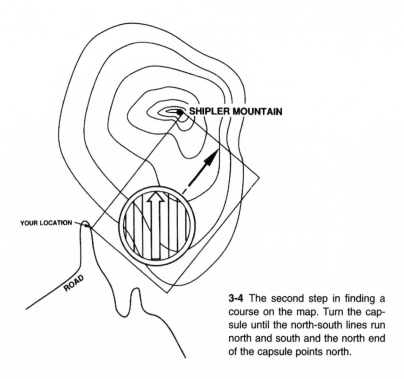

3-4 The second step in finding a course on the map. Turn the capsule until the north-south lines run north and south and the north end of the capsule points north.

right margins of the map. Once again, you're using the compass as a protractor, so the needle's gyrations are irrelevant. Read the course—the angle—at the index mark, where the direction-of-travel arrow abuts the compass dial.

To transfer that angle to the field, pick up your compass and rotate it as a unit, without moving the capsule in relation to the baseplate, until you've placed the needle in the gate. As shown in figure 3-5, the direction-of-travel arrow now points in the direction you want to go. Pick a landmark, walk to it, pick another, and proceed.

There's a simple trick that lets you use your ruler to align the north/south lines more accurately with true north

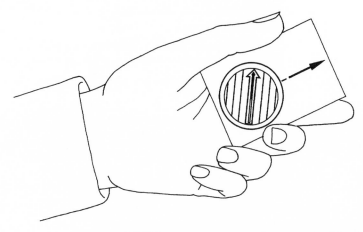

3-5 Using a course measured on a map to determine the direction to your destination in the field. Without moving the capsule in relation to the baseplate, place the needle in the compass gate. The direction-of-travel arrow now points at your destination.

and south. As before, place your compass so one long edge extends between your position and your destination. Butt your ruler up against that edge in such a way that one end of the ruler touches either the right or left edge of the map. Slide your compass along the ruler until the capsule is set over a margin. Now you can use the margin to align the north/south lines accurately. Figure 3-6 illustrates this technique.

The ruler won't always reach the edge of the map, of course, and you need a pretty flat surface to work on. But it's still a useful trick to know when you need high accuracy and conditions permit you to use it.

Now you're pointed in the right direction: you've got a course, taken from the map. But, as three friends and I discovered during the first day of the Colorado Grand Tour, the game's not over. You haven't gotten there yet.

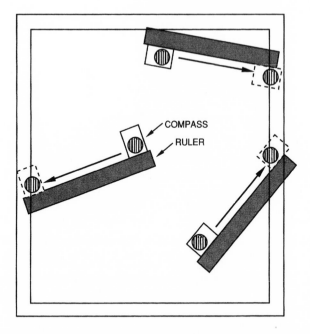

3-6 Using a ruler to align the north-south lines with the edges of the map.

Once again it was stormy, with intense winds and near-zero visibility. (You rarely need a map and compass above timberline when the weather is good. Just pay attention to your surroundings.) My friends were all ex–Outward Bound instructors; I had guided a couple of expeditions on Mt. McKinley. We were bound—we hoped—for Vail, 100 miles and seven days away. But the major storm blasting the Front Range was threatening to thwart us only hours after we started. We lunched at a drafty wilderness cabin, then climbed another quarter of a mile to a knoll clearly marked on the map. From there we needed to follow a broad ridge down through a saddle, then up to the Rollins Pass Road. It didn't seem possible to get lost following a ridge. Further-more, we'd all done this leg of the tour before. Just to be sure, Kim Miller dug out the map and compass and determined our course: 270 degrees, straight west. We weren't going to com-mit the mistake Dale's search party made on James Peak. We pushed on, leaning into the wind like circus clowns with weighted shoes, our heads bowed, our faces so swaddled in face masks and goggles that we felt like astronauts walking on the moon.

A half-hour slipped away—much longer than it should have taken to reach the steep climb leading up to the road. Kim pulled out his compass and called a halt.

"You guys," he said, "we're heading east." We crowded around in disbelief. In the storm, we had walked in a com-plete semicircle and were facing the way we had come.

We righted our course and tried to hurry, but it was still two hours after dark when we arrived in Winter Park, our first night's stop. The moral was clear: when the weather is foul, you not only need to check your compass; you need to check it frequently. Correctly using it once doesn't guarantee safe arrival. Keep your compass in your pocket or around your neck, so you'll have no excuse not to use it.

In the examples above, I assumed you knew your location on the map, but wanted to know the name of a distant peak or the direction you should travel. Now let's assume you don't know your location, but you can identify some landmark.

Two friends and I encountered an easy example of that kind of problem on a Memorial Day hike to Ypsilon Lake in Rocky Mountain National Park. We knew basically where we were; we were on the trail to the lake. But the packs were heavy (we had overnight camping gear as well as ski-moun-taineering equipment) and we wanted to know our exact location on the trail. (What we really wanted to know was how much longer we'd have to suffer beneath those packs.) So I took a bearing on the prominent shoulder of a nearby ridge. Once again I had an angle between a line running north and a line leading to a landmark. I knew that my position had to be along the line of the angle that passed through the ridge's shoulder. So I placed one long edge of the baseplate on the shoulder of the ridge and rotated the whole compass, without turning the capsule in relation to the baseplate, until the north/south lines ran north and south. The direction-of-travel arrow pointed at the landmark. Now I knew I had to be along the line defined by that long edge of the baseplate. I'll call that a line of position. Since I also knew I was on the trail, I had to be at the intersection of my line of position and the trail. Figure 3-7 shows this graphically.

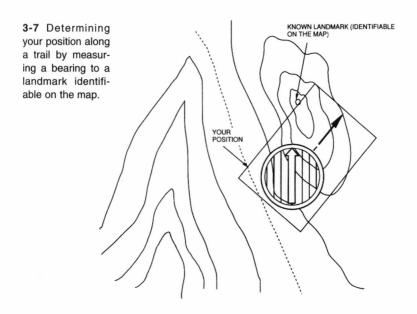

3-7 Determining your position along a trail by measuring a bearing to a landmark identifiable on the map.

KNOWN LANDMARK (IDENTIFIABLE ON THE MAP)

YOUR POSITION

If the edge of the baseplate hadn't reached from the shoulder to the trail, I would have needed a ruler to extend that edge. Sometimes a pencil helps, too, to extend the line of position beyond 12 inches.

You don't need to be on a trail to use this technique. You could just as well be following a stream, river, pronounced ridge or the bank of a large lake. Any kind of prominent, linear terrain feature will do. In all cases, the best accuracy comes from picking a landmark at right angles to the terrain feature you're following.

If you can identify two landmarks, you don't even need to be following a terrain feature. Simply take a bearing off one, and pencil in your line of position on the map. Then take a bearing off the second, and pencil in that line of position. Your location is the intersection of the two lines. Figure 3-8

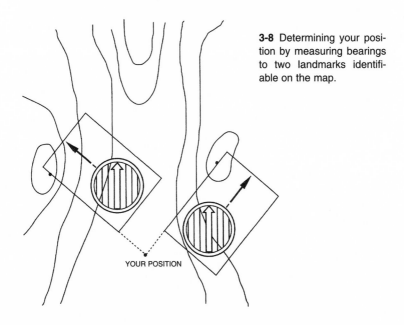

3-8 Determining your position by measuring bearings to two landmarks identifiable on the map.

YOUR POSITION

shows you how. If you can take a bearing off a third landmark and pencil in that line of position, so much the better. Your true position should lie somewhere inside the triangle formed by the three lines of position.

Once you've identified your location with lines of position, perform a reality check. If the lines cross at a stream, and you're standing at a ridge, something is wrong. After we identified our position along the trail to Ypsilon Lake, we took a close look at the map. We had just hiked up a steep section of trail and gained a nearly level stretch. Our line of position from the shoulder crossed the trail right where the contours suddenly became widely spaced—in other words, where the terrain suddenly flattened out. Our location as determined by a line of position checked with the other information we had about the terrain. Whenever you use a map and compass together, it's a good idea to perform that kind of reality check on the results. The key to that is being able to visualize terrain from looking at a map. You can learn a certain amount of that from this book, but to really develop that skill, you need to practice out in the field.

4

True North and the Unfaithful Needle:

HOW TO CORRECT FOR DECLINATION

SOMETIMES READING EXPLANATIONS of how to reconcile true north and magnetic north is like watching a clown cross his arms, point in opposite directions and say, "He went that-away." But fear not: there is a simple, easy-to-comprehend way to solve declination problems. You don't need to memorize rules or silly rhymes; you just need to remember some straightforward logic.

Declination, as I've already mentioned, is the difference in direction between magnetic north and true north. Magnetic north is defined by the direction a compass needle points. True north, also called geographic north, is defined by the direction to the geographic North Pole—one end of the Earth's axis of rotation. Declination, therefore, is an angle, measured with true north as the starting point. If magnetic north lies to the east of true north (to the right, or clockwise as we look at a map), we say the declination is east. If magnetic north lies to the west of true north (to the left, or counter-clockwise), we say the declination is west. East declination is measured clockwise from true north; west declination is measured counter-clockwise. (This is in contrast to bearings and courses, which are always measured clockwise.) In other words, if magnetic north lies 10 degrees east of true north, the declination is 10 degrees east. If magnetic north lies 10 degrees west of true north, the declination is 10 degrees west, not 350 degrees.

Declination in the United States varies from about 20 degrees west in Maine to 21 degrees east in Washington and as much as 30 degrees east in parts of Alaska. The *agonic line,* where the declination is zero, runs from the Great Lakes to Florida. Ignoring the correction for declination can lead you seriously astray and raise the rude possibility of eating shoe leather for dinner as you bask in the warmth of your cigarette lighter. For each degree that your course is in error and each mile that you travel, you'll be off by 92 feet. If the declination is 20 degrees, you'll be off by 1,840 feet—one-third of a mile—after hiking just one mile. It's pretty hard to relocate your tent if a third of a mile of timber or fog separates you from it.

Strictly speaking, it's incorrect to say "the compass needle points to the magnetic north pole." What the needle actually does is align itself with the Earth's magnetic field. That magnetic field resembles a skein of yarn when the cat's done playing with it. Compass needles may or may not actually point at the magnetic north pole itself, which, in 1981, lay just north of Bathurst Island in Canada's Northwest Territories. For reasons not well understood, the magnetic poles move slowly, over periods of many years, through circular paths with a diameter of about 100 miles.

But for a wilderness traveler, all that doesn't matter. If you're using the most recent USGS map available, the declination it gives will be accurate within a degree or two.

Although declination changes slowly as you travel east or west, you can assume that the needle always points in the same direction within the bounds of the area you cover in a typical human-powered trip. On the East Coast, for example, where declination changes relatively quickly, you'd have to travel east or west at least 50 miles for the declination to change one degree. For the sake of simplicity, let's just say the compass needle points to magnetic north.

All USGS maps have a declination diagram in the bottom margin similar to the one shown in figure 4-1, which happens to show a declination of 14 degrees west. True north is always indicated by the line with a star. The magnetic north line is indicated by "MN." On some maps, grid north, a concept we don't need to worry about, is indicated GN. The amount of declination, in degrees, is written beside the diagram.

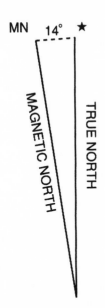

4-1 Approximate mean declination for a map with a declination of 14 degrees west.

Let's assume for the moment that you didn't spend the extra money to buy a set-and-forget compass. Before you can understand how to correct for declination with a standard baseplate compass, you need to ingrain two facts in your mind. First, every angle that you measure on a *map* is measured *clockwise,* with *true north* as the starting point. The needle is irrelevant. You're simply using the compass as a protractor. Second, every angle that you measure in the *field* by placing the needle in the gate of the *compass* is also measured *clockwise*, but the starting point is *magnetic north.*

Let's refer to angles with true north as the starting point as true north bearings or true north courses. (They're essentially the same. A bearing is just a direction to a landmark; a course is a direction you'll follow.) All angles measured on a map, starting from true north, will be true north bearings of true north courses. All angles measured with a compass, using magnetic north as the starting point, will be magnetic north bearings or magnetic north courses.

Now let's assume that you're hiking in the Colorado Rockies, where the declination is always east. Take a look at figure 4-2. True north is marked in zero degrees. Magnetic

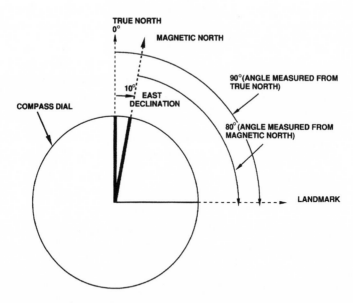

4-2 The relationship between true north angles and magnetic north angles for a declination of 10 degrees east.

north is marked 10 degrees. It lies to the east of true north. In other words, the declination is 10 degrees east. Now let's say you measure a course on the map to a landmark and find it to be 90 degrees. That's a true north course; you started measuring at true north, just as the diagram shows. Now if you take a bearing on that same landmark with your compass, you'll find the bearing to be 80 degrees. That's a magnetic north bearing, since you started measuring at magnetic north.

The true north angle, measured on the map, is greater than the magnetic north angle, measured in the field with the compass by placing the needle in the gate. Furthermore, the difference is 10 degrees—exactly the amount of the declination. And that leads to our first conclusion: *when the declination is east, true north angles (bearings and courses) are always going to be greater than magnetic north angles (bearings and courses).* If you measure an angle on the map and want to transfer it to your compass, you must *subtract* the declination from the true north angle because magnetic north angles are always smaller than true north angles when the declination is east. If you measure an angle with your compass and want to transfer it to the map, you must *add* the declination, because true north angles are always greater than magnetic north angles when the declination is east.

You don't need to memorize these rules. Just remember the logic behind them. If you need to jog your memory, look at the declination diagram on the bottom of the map. Notice how it closely resembles the part of figure 4-2 that is drawn with heavy lines. If magnetic north is east (clockwise or to the right) of true north, then every angle measured clockwise from true north must be greater than the same angle measured clockwise from magnetic north. To reinforce this concept one more time, look at figure 4-3. It shows the relationship between magnetic north and true north for a situation where the true north bearing is 270 degrees.

The same logic applies if the declination is west, as it is on the East Coast. Look at figure 4-4. The declination is again 10 degrees, but this time it's west. Let's say you measure an angle on the map (starting at true north) as 90 degrees. If you measure the same angle in the field with your compass, starting at magnetic north, you'll get 100 degrees. The difference, 10 degrees, is equal to the declination. That leads to our second conclusion: *when the declination is west, true north angles (bearings and courses) are always going to be smaller than magnetic north angles (bearings and courses).*

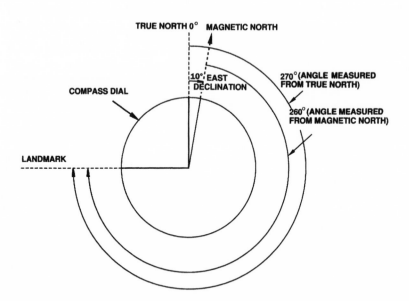

4-3 A second example of the relationship between true north angles and magnetic north angles for a declination of 10 degrees east.

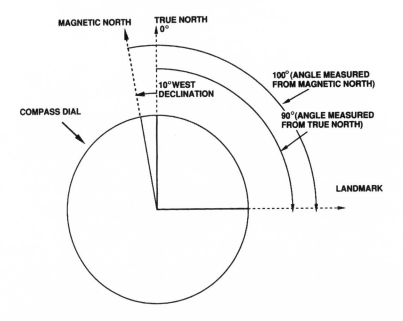

4-4 The relationship between true north angles and magnetic north angles for a declination of 10 degrees west.

If you measure an angle on the map and want to transfer it to your compass, you must *add* the declination to the true north angle because magnetic north angles are always greater than true north angles when the declination is west. If you measure an angle with your compass and want to transfer it to the map, you must *subtract* the declination from the magnetic angle for the same reason: true north angles are always smaller than magnetic north angles when the declination is west. Again, the declination diagram in the bottom margin of the map should help refresh your memory should you ever forget.

Figure 4-5 gives another example of the relation between magnetic and true north when the declination is west, this time for a true north angle of 270 degrees.

Sometimes, in the East, you'll measure a true north angle on the map, then find when you add the west declination that you've gone past 360 degrees. For example, you might measure a true north angle on the map as 355 degrees, then need to add a declination of 10 degrees west to get the magnetic angle: 355 + 10 = 365. An angle of 365 degrees is the same as an angle of 5 degrees. You can also just rotate the compass dial counter-clockwise 10 degrees to add 10 degrees to the true north angle of 355 degrees and get the correct magnetic north angle of 5 degrees. Figure 4-6 shows this graphically.

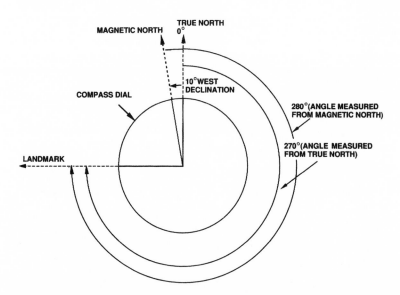

4-5 A second example of the relationship between true north angles and magnetic north angles for a declination of 10 degrees west.

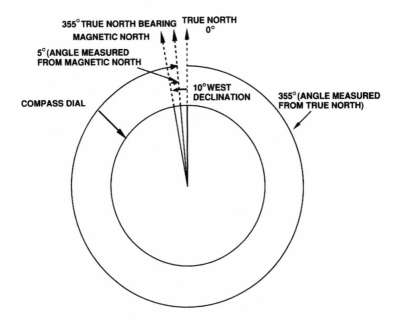

4-6 The relationship between a true north angle of 355 degrees and a magnetic north angle of 5 degrees when the declination is 10 degrees west.

Sometimes, in the West, you'll measure a true north angle on the map, find it's less than the declination, then need to subtract the declination from the true north angle to get the magnetic north angle. For example, the true north angle, measured on the map, might be 5 degrees, and you'll have to subtract a declination of 10 degrees east. Five degrees is the same as 365 degrees, so you can just subtract 10 from 365 and get 355 degrees as the correct magnetic north angle. You can also just rotate the dial clockwise 10 degrees from its original setting of 5 degrees to do the subtraction and reach the correct magnetic north setting of 355 degrees. Figure 4-7 shows this graphically.

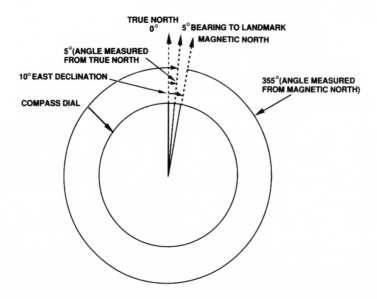

4-7 The relationship between a true north angle of 5 degrees and a magnetic north angle of 355 degrees when the declination is 10 degrees east.

Some people recommend scribing lines on your maps that run toward magnetic north, then aligning the capsule's north/south lines along those magnetic meridians whenever you measure an angle on the map. That way, your starting point for measuring an angle is always the same, whether you measure it on the map or measure it with the compass. Scribing lines is fine if you've got only a few maps to do and a drafting table to do it accurately. To my mind, however, it's much easier to remember the logic of adding and subtracting. Since most people do the majority of their hiking in one area, they only need to remember one line of logic. In the West, true north angles are always greater than magnetic north angles; in the East, true north angles are always smaller than magnetic north angles.

If you're using a set-and-forget compass, you can forget about adding and subtracting once you've set the declination and double-checked that you set it in the right direction. With most such compasses, setting the declination is idiot-proof. Usually you turn a small screw or perform some other simple operation to adjust the compass gate so it points to the angle representing the declination. If the declination is 20 degrees east, for example, the compass gate, after adjustment, would point to 20 degrees. If the declination is 20 degrees west, the compass gate, after adjustment, would point to 340 degrees (360 – 20 = 340).

When you're measuring an angle on the map with a set-and-forget compass, you ignore the needle (as always) and the compass gate (which no longer points to the north on the capsule). Instead, you always use the capsule's north/south lines when you're orienting the capsule north and south, making sure north on the capsule points to north on the map. To transfer that angle to your compass, simply place the needle in the gate. Angles measured with your compass can be transferred directly to the map, again using the north/south lines in the capsule, not the gate. As you can see, set-and-forget compasses let you avoid mental gymnastics when you're cold, wet and would much rather think about dinner courses than compass courses. They also let you avoid spending lots of time drawing lines on your maps. In the time you save, you can easily earn the money to buy the better compass.

5

Bambi Meets Godzilla:

WHEN THEORY MEETS THE REAL WORLD

WHEN WE AWOKE at our camp on the Kahiltna Glacier, menacing clouds had already encircled Mt. McKinley's 20,320-foot summit, two vertical miles above us. By mid-morning, when we started hauling a sled-load of supplies up the glacier, the cloud ceiling had plunged to 10,000 feet and snow was falling so thickly we inhaled dozens of huge flakes at every breath.

In May, the peak climbing season, the Kahiltna Glacier is so heavily traveled by climbers attempting the West Buttress route that a virtual trail leads to the summit. Skis, snowshoes and boots pack a trench into the snow as climbers avoid trail-breaking by following in each other's footsteps. Climbers further mark the trail with bamboo wands flagged with orange surveyor's tape. Heavy storms, however, like the one now assaulting the mountain, knock down and bury the wands and fill in the trail. Landmarks vanish, and a whiteout ensues. Glacier, clouds and mountain become indistinguishable, and all sense of direction vanishes. I've seen climbers in whiteouts step calmly off five-foot ice cliffs, completely oblivious to their existence until they planted a foot firmly in air.

Such a whiteout was quickly overtaking us. We had anticipated that the West Buttress would be crawling with the usual hordes of climbers, their packs sporting orange forests of wands, and so we had brought only a few wands of our own. Now we were regretting that decision. Only a few wands marked the trail behind us, and drifting snow was rapidly filling the track itself. Soon all trace of the route back to our tents would disappear. The tents themselves would be invisible from only 50 yards away. Continuing uphill now might make relocating our camp a few hours later difficult indeed.

I called a halt and urged my teammates to retreat for the day. Reluctantly, they agreed. We cached our loads and bolted for camp.

Fortunately, several teams behind us had beaten down the trail, so finding our tents turned out to be easy. My friends gave me glances that said, "next time, don't be such a wimp." We settled into our sleeping bags to wait for clear weather.

In the morning, when I slipped out of the tent, I saw a line of climbers appearing slowly, one by one, over a low ice hummock well off the correct route. Each step seemed to demand tremendous effort. Their haggard faces told of a sleepless night. Soon their story unfolded. All were from the University of Idaho. They had pushed a little higher than we had the day before, turned back a little later. The trail had vanished, buried by wind-driven fresh snow. When they reached the level glacier only a quarter of a mile from their tents, they lost their way completely. Tired, in failing light, they dug a huge snow cave and waited for dawn, without sleeping bags, foam pads, food or water. The temperature dropped to 15 degrees.

Foresight would have prevented a very unpleasant night. On glaciers and big snowfields, where whiteouts can sweep in quickly, it's often essential to create your own landmarks by placing wands every 180 feet or so. Climbers should always travel roped up on a big glacier as protection against falling into a crevasse. Most teams use a 150-foot rope. If the wands are spaced every 180 feet, the second climber can stay at a wand while the leader goes out to find the next. That way the team is never out of sight of a wand.

Once the Idaho climbers lost the trail, they had few options. Bivouacking in a snow cave, out of the wind, where they could share body heat, probably prevented any serious hypothermia. If they had been determined to keep searching, their best bet would probably have been to backtrack, using a course taken off the map, until they reached the first steep rise—proof positive that they were above their tents. Then they could have zig-zagged slowly up the slope, from one side of the glacier to the other, until they found one of the remaining wands. From there they could have measured a course on the map straight down the glacier. Then, after fanning out but remaining in sight of each other, they could have started searching, following the bearing downhill. The last person on the lead rope would carry the compass and call directions to the leader, just as we did on Mt. Hunter. The methods certainly wouldn't guarantee finding their tents; but it would at least keep them moving and warm until morning.

Another glacier 8,500 miles away, in Argentina, put me in a predicament similar to the Idahoans'. The experience taught me a lot about navigating in the real world.

I was making a one-day, solo attempt on the summit of Aconcagua, the highest mountain in the Western Hemisphere at 22,834 feet. I had started at 19,000 feet, at the foot of the Polish Glacier, at 2 a.m. Thirteen hours later I was still 500 feet below the summit. A flu virus had stolen my voice, my muscles had become lard and a flotilla of black clouds was rolling in from the Pacific. I headed down.

And none too soon. By the time I reached the broad snow bowl at the foot of the glacier, it was nearly dark. A few wands materialized out of the gloom, left by some previous expedition. They seemed to be leading too far to the right, or south, but I followed them anyway. I knew that just beyond the foot of the glacier was the top of a large cliff. My camp lay below the cliff; I had walked around the north end of the cliff in the pre-dawn darkness 16 hours before. The cliff-top would act as a *catching feature,* alerting me that I needed to turn left, to the north, to begin my end-run around the cliff.

I walked off snow onto rock and stumbled ahead. Suddenly an abyss yawned before me out of the darkness and flying snow. I had reached the cliff-top. The crux now was to find my way back through the complex, broken cliff bands that formed the north end of the cliff.

I began scrambling northward along the junction of snow and rock, thinking I was only minutes from my tent. But nothing looked familiar. The cliff shrank and ended but now I confronted a talus field studded with outcrops and small cliff bands. In darkness and storm, dehydrated and exhausted, I could not find the way back to my tent. I croaked "hello, anybody there?" a couple of times, but got no reply. I started thinking about digging a snow cave and waiting for dawn.

Then, barely audible over the wind, I heard a voice. I stumbled in that direction. A tent appeared, inhabited by some American climbers. They pointed me in the right direction, at last, and I soon found my tent only 300 yards away. Too exhausted to eat the dinner my body desperately needed, I drank a quart of soup and collapsed into my bag.

I had made a serious mistake that had almost cost me a forced bivouac on a sub-zero night at 19,000 feet. I hadn't built a cairn, or series of cairns, to guide me around the cliff. After all, the weather had been perfectly clear when I started, and I expected to be back before dark—the oft-repeated refrain of lost hikers and climbers everywhere. In situations like that, it's best to plan for the worst case, not the best.

But I had done some things right. I had used a technique called "aiming off." Instead of aiming directly for the end of the cliff, I had aimed to the right. Then, when I reached the edge of the glacier, I knew I had to turn left. If I had aimed directly for the cliff end and missed even slightly, I would have wasted even more time than I did wandering in the dark amidst the cliff bands and outcrops forming the cliff's indistinct end, wondering whether to go right or left.

You can apply the same technique to following a bearing through the woods back to a road where you parked your car. It's impossible to follow a bearing with complete accuracy. If you miss by even a couple of hundred yards, your car may be hidden by a bend in the road and you won't know which way to turn. So, instead of aiming directly for the car, set your course about 10 degrees off. Then, when you hit the road, you know which way to turn.

The same principle applies in other situations: finding a bridge or ford across a creek, a snow bridge across a lengthy crevasse, a camp you've placed along the shore of a large lake. Road, creek, crevasse and lake shore are all "catching features": they tell you unmistakably that it's time to change course. Once you've made the turn, they can be considered *handrails.* You can follow them without further reference to your map or compass. Thinking about those two concepts can often make your route-finding easier. Instead of heading cross-country for several miles, navigating through thick woods with a compass, it may well be easier to walk an extra quarter-mile to a stream or lakeshore that parallels your course and serves as a handrail. It will almost certainly be easier to walk that far to a trail rather than bushwhack. Then look for a catching feature to tell you when to resume your original course. It might be a prominent side stream. It could also be a particular bearing on a prominent peak that you can see from the handrail.

It's all too easy, when you're traveling through the woods, to decide you've reached the feature you want to use as your handrail when you actually haven't. In fact, it's remarkably easy, if you're careless or in a hurry, to make the map seem to fit what you're looking at in any situation. Joe Kaelin and I demonstrated that perfectly in 1979 in the Canadian Rockies.

Admittedly, the map was poor: a 1:250,000 scale topo on which one inch equaled four miles. And the clouds were hanging only a few hundred feet above the valley floor. Our intent was to cross the Sunwapta River and hike up Habel Creek to Wooley Shoulder. We made a crude estimate of the distance we should drive from Sunwapta Pass to reach the junction of Habel Creek and Sunwapta. When we'd driven about that far, we spotted a side creek and told the driver who had given us a lift to drop us off.

After three hours of hard work and 2,000 feet of elevation gain, the fog lifted from both the terrain and our minds. We realized we were in entirely the wrong drainage. Habel Creek intersected the Sunwapta a mile further down-river. We descended almost 2,000 feet, reached the right creek, and climbed 3,000 feet to Wooley Shoulder.

Even with our small-scale map, we should have done much better. Habel Creek clearly occupied a deep valley, much larger and much less precipitous than the one we'd so laboriously climbed. Any kind of careful map study would have told us very quickly that we were gaining elevation much too fast and were in much too shallow a ravine to be in the Habel Creek valley. More careful measurement of the distance from Sunwapta Pass to our drop-off point would also have helped us get started correctly.

A couple of years later and wiser, I encountered a similar situation on my trip to Aconcagua. Our map there had 500 *meter* contour intervals. Once again we were traveling along a major river, the Rio de las Vacas, and were looking for a side canyon occupied by a stream called the Relincho. The Relincho, we knew, drained the entire eastern side of Aconcagua, including the Polish Glacier. We passed several side canyons, including one the assistant guide thought for sure contained the Relincho. But none of the side canyons seemed to me big enough to contain the outflow of a major glacier. A few mile further up the Vacas, we turned a corner and found ourselves gazing up a big side valley at Aconcagua rising two vertical miles above us.

As my experience on Aconcagua shows, sometimes map and compass alone won't get you there. You have to use your head, too.

That's good advice as well for correcting route-finding mistakes that have left you temporarily confused, shall we say, about your location. If you've been locating your position on the map periodically, and kept track of the direction you've been traveling, you can't really become lost. All you need to do is backtrack to your last known location. Since there's always a possibility that you will need to reverse course, for many reasons besides getting lost, you should glance over your shoulder frequently to memorize what your route looks like when you're heading in the opposite direction.

If you've been following a bearing and think you should have reached your destination by now, but haven't, stop and analyze possible mistakes. Did you compensate for declination? Did you compensate in the right direction? Could you have overshot your destination? Or are you just moving more slowly than you thought, and it's still ahead? Often it pays to go to a nearby clearing or bare-topped knoll. The clear view may help you identify landmarks. Whatever you do, don't panic, and don't blunder off in some hastily chosen direction, compounding your confusion.

Following a stream downhill is *sometimes* good advice, if you're in a pretty civilized area well cut by roads. In real wilderness, however, you could find yourself walking a long, long way. In Canyonlands National Park, for example, following a stream downhill could take you to the Colorado River. No road parallels the river; there isn't even a bridge *across* the river for a hundred miles.

If you're absolutely convinced you cannot determine your location, and if you notified someone where you were going and when you expected to be back, your best bet is to stay put and wait for searchers to find you. Nearly all wild areas in this country are within the response area of some kind of search-and-rescue organization. You'll make their job easier if you can move to a nearby location that's easily visible from the air and the surrounding terrain—a ridgetop or some kind of clearing. Building a small fire—*if* you're *utterly* convinced you can keep it under control—makes you easier to spot and will certainly keep you warmer inside and out.

But all this doom-and-gloom stuff should happen to the other guy, not to you, if you find yourself on the map at the beginning of your trip, pinpoint your position periodically, and always know the direction you're traveling.

6

Weighing Air to Find Your Way:

ALTIMETER NAVIGATION

A POWERFUL STORM punched into the Rockies the night before three friends and I embarked on the Colorado Grand Tour, the seven-day, 100-mile ski-tour from Eldora to Vail that I mentioned in chapter three. All the sound mountaineering judgment we possessed dictated postponing the trip until the storm ebbed, for our route touched the Continental Divide four times and lay above timberline for much of the way. But we had scheduled every night's lodging for a particular date; postponing would have inconvenienced our waiting friends. We plunged into the blizzard, armored like modern knights in full winter regalia: complete wind suits, gloves with gauntlets reaching to our elbows, hoods, face masks and goggles.

For six hours we skied high above timberline on a nearly featureless, undulating plateau. When the wind gusted and kicked up a ground blizzard, my companions would vanish, although they were only 50 feet away. In theory, we were following the Rollins Pass road, but drifted snow buried it almost everywhere. We knew from the map that we would soon have to make a sharp left turn—but where?

A small sign bearing no marks but an arrow pointing left materialized out of the storm. Could this be the place? Our compasses were useless in pinpointing our location because there were no landmarks to sight. But Peter was carrying an altimeter. It read 11,670 feet—exactly the same elevation as the pass, according to the map.

We took a bearing off the map (assuming we were indeed at the pass) and turned left. An hour later, the clouds lifted briefly. Ahead lay Winter Park, our first day's destination. Soon we were floating ecstatically down a ravine brimming with powder snow. Two hours after that, we were wolfing spaghetti and beer in the welcome warmth of a friend's home in Winter Park.

An altimeter gives you information about your location even when a compass is useless except to take a bearing off the map. One glance at the altimeter gives you the elevation of the contour line you're on. If you have a second line of position—you know you're on a particular trail, for example, or following a particular stream or ridge crest—then your position is pinpointed exactly. You're standing at the intersection of the contour line and the second line of position. Sometimes, particularly if you're following a trail or ridge, your contour line will intersect twice. In that case, having even the vaguest idea of where you are (which side of the pass you're on, for example), should tell you your location exactly.

Figure 6-1 shows two examples. If your altimeter reads 10,000 feet, and you know you're on the trail to Chasm Lake, then you know you're at point A, where the 10,000-foot contour line intersects the trail. Since that contour line intersects the trail only once, there can be no confusion. If it intersected twice, you would have to determine which of the two

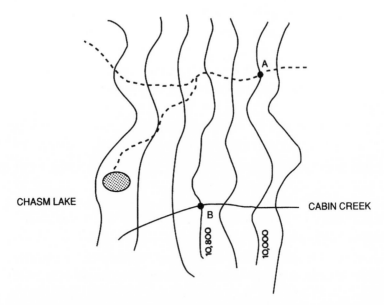

6-1 Using an altimeter and another line of position (such as a trail or creek) to determine your position.

intersections marked your position by other means: the general lay of the land, the time spent on the trail and your estimate of your pace.

In the second example, let's say you're bushwhacking up Cabin Creek and your altimeter reads 10,800 feet. You know you're at point B, where the creek intersects that contour. The same contour line never intersects a stream twice, of course, because streams constantly flow downhill.

The second line of position can also be a compass bearing. Perhaps you're in thick woods and get a glimpse of an identifiable peak. Or perhaps you're above timberline in a storm and get a glimpse through a hole in the clouds of some identifiable pass. With a little luck, the compass bearing will cross the contour line indicated by your altimeter at something close to a right angle. If the bearing runs parallel to your contour line, all you've done is confirm your altimeter's reading. You can't pinpoint your position.

Mountaineers often find altimeters valuable even when a compass is useless and a map serves only to indicate the elevation of the top and bottom of the climb. On a first ascent on the north face of Alaska's Mt. Foraker in 1983, we used an altimeter to keep track of our progress up a steep rib. Since we were making very little horizontal distance on the map, compass bearings couldn't pinpoint our position. Our altimeter, though, let us gauge our rate of ascent and ration our food accordingly. You can use the same principle to judge your progress on a trail. If you know from the map that the lake is at 9,000 feet and the pass is at 12,000, you can take note of the time when you leave the lake and estimate your rate of ascent and time of arrival by how long it takes to climb the first 500 or 1,000 feet. If you leave the lake at 8 a.m. and take an hour to climb the first 1,000 feet, you can estimate you'll take three hours to climb 3,000 feet, which will put you at the pass at 11 a.m. In reality, with rest stops and fatigue, you'll probably arrive a little later.

In some situations, your compass and your altimeter can pinpoint your location even in a whiteout. The key to this direction-of-slope method is using the compass to determine the bearing of an imaginary line pointing straight downhill. Skiers would call it the fall line. By definition, that line will be perpendicular to the contour line (determined by your altimeter) on which your position lies. Take a look at figure 6-2. If your altimeter tells you you're on the 10,200-foot contour, for example, and your direction-of-slope line runs at a bearing of 150 degrees—southeast—you know you must be at position A. If the bearing is 210 degrees, you know your location must be position B.

This method won't help you everywhere, obviously. You can't determine your position on a broad hillside resembling a tilted tabletop, for example. But knowledge of this method would have helped prevent three friends and I from getting lost on the Commando Run, a ski tour from Vail Pass to Vail, Colorado. The dotted line on figure 6-2 shows the route we should have taken. Instead of climbing the hill there, however, we continued contouring at the 10,200-foot level until we reached point B. When we finally turned uphill, we found ourselves forced to switchback repeatedly, an exhausting and tedious process, then cross a steep and avalanche-prone chute at the very top of the slope. A quick map check an hour earlier would have saved a lot of energy.

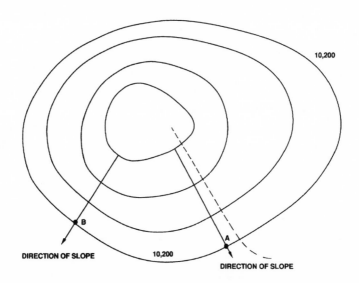

6-2 Using an altimeter and the direction-of-slope method to determine your position.

In all of the techniques I've just described, the accuracy of your position estimate depends on the accuracy of your altimeter. Altimeters vary widely in accuracy, with really good ones costing $190 or so. The accuracy of an altimeter also depends, to a degree, on the user. To maximize your altimeter's usefulness and correct possible errors, you need to know a little about how altimeters work.

Altimeters are closely related to barometers. Both work by measuring air pressure: in effect, the weight of a column of air rising above the instrument's position. The higher you go, the less atmosphere lies above you. Altimeters translate that pressure drop into an altitude reading.

Pocket altimeters use a small, sealed metal capsule to sense air pressure. Most of the air inside is removed during manufacturing, so a partial vacuum exists. As the pressure outside the capsule varies, the walls of the capsule flex in and out. Those movements, which are only a few thousandths of an inch, are translated via a complex mechanism to the movements of a pointer, which rotates around a dial to indicate the elevation. The dial can be adjusted in relation to the pointer, which allows you to correct errors by setting the altimeter to the proper elevation. If you arrive at a lake that the map tells you is at 3,000 feet, for example, and your altimeter tells you it's at 3,100 feet, you can adjust the altimeter back down to 3,000. On most altimeters, the pointer also indicates the pressure directly, so the instrument functions as a barometer.

All metals expand and contract as the temperature fluctuates. Inexpensive altimeters can't compensate for that, which means that a change in their temperature causes them to register an apparent change in altitude even when the instrument is stationary and the pressure unvarying. Using an inexpensive altimeter that's not temperature-compensated can lead to errors of as much as 600 feet if you take the instrument out of a warm pocket and let it chill to ambient temperature on a wintry day. To minimize such errors, let the altimeter adjust to the prevailing temperature before setting it to your starting altitude. Then keep the altimeter in an outside pocket of your pack, so it remains at approximately the same temperature.

Better instruments incorporate some element in their design that counteracts the effect of instrument temperature. It's usually a bimetallic strip made of two metals fused together, that have widely different rates of expansion and contraction when heated or cooled. When the strip is heated, for example, the strip bends away from the metal that expands the fastest. That bending is used to counteract expansion in other parts of the instrument that would otherwise lead to an apparent change in altitude.

The first source of error, therefore, can be greatly reduced simply by buying a high-quality, temperature-compensated instrument. The second source of error must be controlled by the user.

The second error arises because air pressure fluctuates constantly, even at the same elevation, as the weather changes. If you're moving at the same time, it's impossible to sort out the two influences on your altimeter. The only solution is to reset your altimeter at known elevation points—a pass, a lake, a summit, the point where a trail crosses a stream or cuts across a prominent ridge. If you can reset your altimeter every hour or so, you can limit the error due to weather to 40 feet or less—usually less.

Larger errors due to weather are common when you camp somewhere for a day or longer. If there is a major weather change during the night, for example, the difference between evening and morning readings can easily be several hundred feet. If you know the exact elevation of your camp from the map, you can correct as you normally would. If you can't determine your elevation from the map, the best way to maintain the accuracy of your altimeter readings is to note the reading when you arrive in camp, then reset the altimeter to that reading when you leave.

The easiest way to correct for both possible errors in your altimeter is simply to reset it at known elevation points as frequently as possible. With less effort than it takes to sight a single landmark and plot the bearing on the map, your altimeter will then serve you well as a navigational tool.

7

Never Get Lost Again?

HOW THE GLOBAL POSITIONING SYSTEM IS REVOLUTIONIZING BACKCOUNTRY NAVIGATION

THE GLOBAL POSITIONING SYSTEM (GPS) is the most significant advance in navigation since the invention of the compass in the 12th century. Today, by carrying a handheld device weighing less than a pound and costing less than a good pair of boots, a wilderness traveler can know his position within 100 meters anyplace in the world, day or night, regardless of the weather.

Fishermen are using GPS receivers to find prime fishing holes located far from any trail. Cross-country skiers are using them to relocate their cars, even when drifting snow has obscured their tracks and low clouds are hiding all landmarks. Backpackers are using them to relocate their tents at night after lingering on some high overlook to watch the sunset. And mountaineers are using them to navigate across treacherous glaciers when sudden storms produce blinding whiteouts, in which even the ground underfoot seems shifting and uncertain.

Boosters argue that the use of GPS receivers means that nobody should ever get lost again. As astounding as their capabilities are, however, GPS receivers are not infallible substitutes for good map-reading skills, a compass and common sense. To get the most out of them, it is essential to understand their tremendous strengths—and real weaknesses.

The heart of the Global Positioning System is a constellation of 24 satellites that orbit the Earth twice a day. These satellites were originally launched by the U.S. military to provide a navigation system for its troops, ships and planes. Each satellite continuously broadcasts its position and the exact time. At least five satellites are above the horizon at all times, regardless of your position on Earth. The time required for a signal to reach the GPS receiver indicates the distance between the receiver and the satellite. By combining the information received from three or more satellites, a GPS receiver can calculate your exact position in degrees of latitude and longitude. Take that information to a map and you'll know where you are.

Simple position fixes are only the beginning of a GPS receiver's capabilities. GPS receivers also let you name and record the position of landmarks. You can record landmarks as you reach them by turning on the receiver, letting it get a position fix, then entering that position into the GPS receiver's memory. You can also record landmarks by measuring on a map the latitude and longitude of a landmark, then entering those coordinates into the receiver along with a name.

Almost all GPS receivers include a "go to" feature. Just press the "go to" button, select a landmark from the list that appears on the receiver's LCD display, and the receiver will tell you the distance to the landmark and the compass bearing you must follow to reach it.

Let's say a foot of fresh powder snow has just blanketed the high country and there's more in the air. You know the new snow means there will be some great skiing above

Powder Lake, but you also know that the fresh snow will have erased all traces of the Powder Lake trail. With a GPS receiver, that's no problem. First you use a map to calculate the position of Powder Lake (more on that below). Then you enter that information in the receiver as a landmark named *powdlk*. Next you drive to the trailhead, let the receiver get a position fix and enter that as a landmark named *truck*. Then you press "go to" *powdlk*. The receiver tells you the lake is three miles away at a bearing of 242 degrees. You ski through the woods to the lake, cut up the powder until your thighs are jelly, then go exploring north of the lake for several hours. By now the continuing storm has covered your tracks again. How do you get back to your truck? Just press "go to" *truck* and the receiver will tell you the distance and compass bearing you must follow to reach your vehicle.

GPS receivers also allow you to create "routes" when it's impracticable to make a direct beeline towards your destination. For example, you could record the position of your truck, then record a series of landmarks as you head out for a day of ski touring. Logical landmarks might include a major stream crossing, a lake, a small meadow with a distinctive dead snag, and a saddle a half mile below the summit. To find your way back to your car after lunch, you can create a route from your position (the summit) through each landmark in turn, ending at your truck. The GPS receiver will guide you by providing the distance and compass bearing to the next landmark. When you reach a landmark, the receiver automatically switches to the next landmark in sequence, providing the distance and bearing to it until you reach your vehicle again. You can also create a route at home by entering the coordinates of the various landmarks you expect to pass on the way to your destination.

The easiest way to determine the position of a landmark is always to go there and let the receiver do the work for you. In some situations, however, you'll want to determine the coordinates of a landmark by measuring them on the map. You will especially want to do this when your destination is a place you've never visited before. Once you've entered the coordinates in your GPS receiver as a landmark, you can press "go to," select a landmark and know exactly how to reach it.

Measuring the exact latitude and longitude of a landmark on a map is a bit more tricky than it sounds. Here's the easiest way: Start by using a ruler or other straightedge to draw a line from the landmark to the right or left edge of the map (whichever is closest). We'll call this the *latitude line*. Draw a second line to the top or bottom edge of the map. We'll call this the *longitude line*. The lines should be perpendicular to the edges of the map. Most maps have the latitude and longitude marked at intervals along the borders. By noting where your latitude and longitude lines intersect the borders of your map, you can eyeball an approximate set of coordinates for your landmark.

To determine these coordinates accurately, however, you have to figure out a latitude scale: the number of minutes of latitude that corresponds to one inch on the map. For example, let's say that the distance between 39 degrees, 52 minutes north latitude and 39 degrees, 53 minutes north latitude is two inches on your map. That's two inches per minute of latitude, or one inch for 0.5 minutes. If the latitude line of your landmark is one inch north of the 54-minute mark on the border of your map, its latitude is 39 degrees, 54.5 minutes. On a 7½-minute USGS quad, one inch on the map equals 0.33 minutes of latitude.

To determine the longitude accurately, you have to figure out a second scale for the number of minutes of longitude that correspond to one inch on that particular map. The latitude and longitude scales are not the same because the distance between two lines of longitude varies depending on your latitude. (Remember that longitude lines converge at the poles; if you're in the Arctic, one degree of longitude is a much smaller distance than if you're at the equator.) Once established, the longitude scale remains the same for maps at the same latitude, but would not be the same for maps of areas a few hundred miles north or south. Figuring out the exact longitude and latitude of a point on the map is most easily done in the warmth of your home, with a flat place to lay out the map, a yardstick, a pocket calculator and plenty of head-scratching time.

It's easier to calculate the coordinates of a landmark when using Universal Transverse Mercator (UTM) coordinates, which most GPS receivers can handle. Before you throw up your hands in horror at the thought of learning a whole new coordinate system, read the next few paragraphs. The system is simpler than its arcane name implies, and in fact is more intuitive than latitude and longitude for use in the field.

The UTM coordinate system is a metric grid system covering all of the Earth except areas near the poles. Its basic unit is a long, narrow strip of the Earth called a zone. It takes 60 zones to cover the entire Earth. Zones are numbered 1 through 60, starting at the 180th meridian and counting eastward. Your position in UTM coordinates is defined first by the zone you're in. Your position inside each zone is then defined by your distance, in meters, north of the equator (your "northing") and your distance east of the zero line (your "easting"), which is defined as a line 500,000 meters west of the central meridian of the zone. For our purposes, you can think of your easting as the distance east of the western boundary of your zone. For example, the UTM coordinates of my home in Boulder, Colorado, are 13 4 **78** 075 E and 44 **25** 340 N. That means I'm in zone 13, at a point 478,075 meters east of the western boundary of the zone and 4,425,340 meters north of the equator.

The legend on most large- and intermediate-scale USGS topos will tell you your UTM zone. Along the borders of the map, you'll find short blue lines set at 1,000-meter intervals. These blue lines are referred to as *1,000-meter grid ticks.* Each tick on the right and left sides of the map is labeled with the northing of that point, expressed in meters. Each tick on the top and bottom sides of the map is labeled with the easting of that point, again expressed in meters.

For the sake of simplicity in reading the map, the last three zeros of the number are eliminated except in the tick marks near the bottom right and top left corners of the map. To further simplify map reading, the first one or two numbers of the northing and easting are printed in small type, as typically they are a constant over one map sheet. The grid tick labeled $_4$74 on the bottom edge of the map, for instance, is 474,000 meters east of the western boundary of your zone. The grid tick labeled $_{37}$19 on the right edge of the map is 3,719,000 meters north of the equator. You'll notice that all I've done to convert an abbreviated grid tick number to meters is add three zeros and place the commas in the appropriate places. Note also that a difference of 1 between two abbreviated grid tick numbers equals 1,000 meters. The distance between grid ticks $_4$74 and $_4$75, for example, is 1,000 meters. Beware: many of the popular *Trails Illustrated* topos do not show the UTM grid ticks.

One beauty of the UTM system is that the units are intelligible. If someone told you that you had to walk 2 minutes of latitude to the north to reach your car, you'd probably have no idea if you would be home in an hour or a week. (You're actually just over a mile from home.) On the other hand, if someone tells you that you have to walk 1,000 meters (one kilometer or about six-tenths of a mile) you know you probably should be there in less than 30 minutes. If you set your GPS receiver to give you your position in UTM coordinates, you can watch the meter turn (literally) as you walk. Your position, as expressed in meters, will change with every few strides. Another beauty of the UTM system is that the scale is always the same for both northing and easting.

The first step in measuring the UTM coordinates of a landmark is to draw lines connecting the grid ticks on the map borders so that you have a visible UTM grid. Using the ruler on the edge of your compass, measure the distance from your landmark to the closest grid tick line to the west. Transfer that measurement to the kilometer scale at the bottom of the map. A kilometer is 1,000 meters; smaller divisions on the map scale represent 100 meters. On a 1:24,000 scale map (the scale used on standard 7½-minute USGS quads), one inch equals 600 meters. If your landmark is a half inch east of the nearest grid tick line, you know to add 300 meters to the easting of that line. Use the same technique to determine the northing of your landmark, and you know your landmark's UTM coordinates.

At least one company sells a handy UTM measuring device that contains three transparent sheets of plastic, each with a UTM grid marked in black. Each sheet matches one of the three most commonly used map scales: 1:24,000, 1:25,000 and 1:63,360. One square on the grid represents 100 meters.

You use the transparent UTM grid to determine the distance, in meters, from the first grid tick line to the west and the first grid tick line to the south. For example, let's say the nearest line to the west is labeled $_2$86, meaning that line is 286,000 meters east of the western boundary of your zone. The nearest line to the south is labeled $_{41}$57, meaning that line is 4,157,000 meters north of the equator. By placing the bottom left corner of the transparent card's grid on the intersection of those two lines, you can see that the position of the landmark you're measuring is 700 meters east and 300 meters north of the intersection. In the UTM coordinate system, that puts you at 13 3 86 700 east, 41 56 300 north—zone 13, 286,700 meters east of the western boundary of your zone and 4,157,300 meters north of the equator.

Once you've determined the UTM coordinates of your destination, you can enter them into your GPS receiver as a landmark. Press "go to," select the new landmark you just created and the GPS receiver will give you the distance and bearing of your destination.

The same techniques apply when your receiver has given you your position and you need to plot that position on your map to see where you are. Some higher-end receivers shortcut this laborious process by giving you your position in relation to a reference point that you define on the map. For instance, you could enter the coordinates of the southeast corner of the map as your reference point. Since those coordinates are printed on the map, you don't need to do any fancy calculations. The GPS receiver then gives your position as, for example, three inches west and four inches north of that corner of the map.

All the fanfare about GPS receivers makes it sound like your compass and altimeter have just become obsolete.

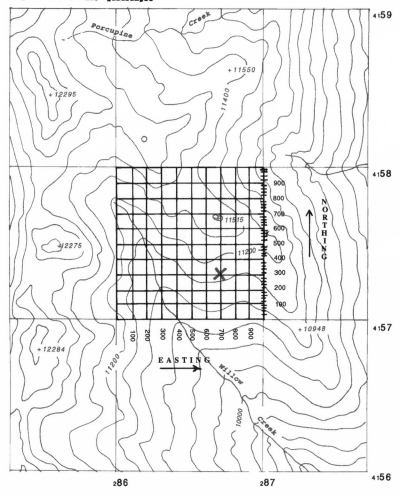

7-1 Let's say you want to know the UTM coordinates of the X marked on the map so you can enter them into your GPS receiver. First, look at the bottom of the map to determine the easting of the first UTM grid line west of the X. It's marked $_2$86. Then look at the right margin of the map to determine the northing of the first UTM grid line south of the X. It's marked $_{41}$57. Then position the UTM card with the lower left corner of the grid on the intersection of those two UTM grid lines, as shown. The X is 700 meters east of the $_2$86 grid line and 300 meters north of the $_{41}$57 grid line. You know from the legend on the map that this quadrangle lies within zone 13. The UTM coordinates of the X are 13 2 86 700 E, 41 57 300 N—zone 13, 286,700 meters east of the western boundary of the zone; 4,157,300 meters north of the equator.

Before you send them to the Smithsonian as national heir-looms, however, there are a few more things you need to know. GPS receivers run on batteries; if the batteries die, and you have no replacements, you're lost. Good GPS receivers are quite shock- and water-resistant, but any electronic device can break. Cliffs and dense foliage will block the signals from the GPS satellites. If you're in thick woods, you'll probably need to find a clearing in order to get a position fix. If you're in the depths of a slot canyon in southern Utah, you may have to climb out of the canyon to determine your position.

Today's civilian GPS receivers *could* give you your position to within 25 meters horizontally—if the military chose to let them. At present, however, such accuracy has been deemed dangerous, as it would provide potential enemies with too precise a tool. Accordingly, the military introduces deliberate errors into the signals from the satellites, which reduce the accuracy to 100 meters 95 percent of the time. Five percent of the time, the error may be larger. In government-speak, these errors are known as selective availability. GPS receivers also provide you with a very rough altitude (plus or minus 150 meters). A good pocket altimeter, if reset frequent-ly at points of known elevation, is much more accurate.

One hundred meters horizontally doesn't sound like a very big error—until you try to find the heavy camera bag you stashed in the middle of a forest while you scrambled up a peak. That 100-meter inaccuracy is also significant as you close in on a landmark. At distances of a quarter mile or more from the landmark, the bearing provided by the GPS receiv-er is pretty accurate. At distances of less than a quarter mile, the bearing can be off by a large margin. My advice: Note the bearing provided by the device while still a quarter mile away from your landmark. Transfer that bearing to a good compass, and follow the compass needle for the last quarter mile until you reach your destination. The distance to the landmark, provided by the GPS receiver, seems to be considerably more accurate right up to the point where you reach the landmark.

To make your life easier still, try to select landmarks that are easily visible from some distance away. A lone clump of trees above timberline or a distinctive rock outcrop in the middle of a large meadow makes a good landmark. Use your common sense: My GPS receiver, for example, tells me I live on the north side of Table Mesa Drive (when I plot the coordinates it provides for my house on a topo map), but I know for a fact I live on the south side.

GPS receivers are not difficult to learn to use, but mastering them does take a little practice. Be prepared to spend several hours working with a new receiver before you take it into the field. If you haven't used your receiver since last summer, throw in the instruction book as well.

Do you really need a GPS receiver? In the summer, probably not, especially if you stick to well-marked trails. During this time of year, a map and compass and, perhaps, an altimeter are the only navigational tools you need. If you like to travel cross-country in the summer, you may find a GPS receiver handy once in a while. Remember, however, that summertime cross-country navigating is usually most difficult below timberline, when trees obscure your view of landmarks—a situation in which GPS receivers are less than completely reliable.

If you do a lot of winter travel in the snowy ranges, or travel at any time of year in glaciated ranges, you may find a GPS receiver valuable indeed. If the people on the hapless University of Idaho expedition to Mt. McKinley that I described in chapter 5 had had a GPS receiver, they could have easily navigated back to their tents regardless of the weather and spent a comfortable night there instead of spending it shivering and hungry in a snow cave. As I review my own backcountry career and those moments when I was most worried about getting lost, I can see that a GPS receiver would have been well worth the cost and the extra 10 ounces or so in my pack. As the accuracy of GPS receivers improves, adventurous wilderness travelers may come to regard them as essential.

8

Running With Cunning:

THE SPORT OF ORIENTEERING

SOME PEOPLE NO SOONER MASTER a skill than they feel a need
to test their mastery in competition with others. When people of that type combined map and compass skills with cross-
country running, the sport of orienteering was born.

It began in Sweden in the 1920s, then sprouted in the
United States in 1946. Today U.S. orienteering clubs stage
over 450 local meets each year, drawing over 20,000 people.
Competitors gather for another 50 regional and national
events as well. In Sweden, Norway and other European coun-
tries, orienteering is a national pastime. A single big meet in
Sweden can pull in as many as 17,000 enthusiasts.

In the most popular form of orienteering, competitors
must find, in correct order, a series of five to 15 control points
hidden in the woods. A red and white, prism-shaped struc-
ture made of fabric or cardboard marks the control point. The
controls, as they are called, are either marked on the map
furnished to each participant at the start of the race, or
copied by the participant from a master map after the starting gun sounds. Easy courses might extend for 10. The win-
ner is the runner who finishes in the shortest time after

locating all the controls. Runners who miss a control are disqualified.

Orienteering maps are usually drawn to a larger scale than standard USGS topos. A scale of 1:10,000 is common. The top of the map always represents magnetic north instead of geographic north, so there's no need to worry about correcting for declination. Meet officials field-check each map as they plot the course and place the controls. They delete trails that have vanished, mark any new ones and add the details that are one key to finding the controls: fences, boulders, knolls, tiny streams that would escape the USGS's notice. In addition to the map, runners get a very brief description of the terrain feature at which each control point will be found.

Orienteering races always use a staggered start, so runners can't simply follow each other from control to control. Ideally, the controls are also placed so runners approaching a particular control get no clues from others leaving it.

Only in beginner's races will a straight-line compass course ever be the fastest way from one control to the next. Victory doesn't necessarily go to the fastest runner. Instead, it goes to the competitor who can visualize the best route from a rapid study of the map, then follow it quickly and accurately. Expert orienteers use many of the techniques I've already described, which in fact originated in the sport. Wherever possible, for example, they look for handrails that lead them in the right direction. Trails and roads are the most obvious handrails; power lines, fences, streams and edges of fields are less obvious but equally effective. Over a short distance, the sun can be a handrail. Perhaps you can follow your shadow, or run with the sun full in your face. Or perhaps you can tell from the map that you need to follow a course that crosses the shadows thrown by trees at a 90-

degree angle. With a good handrail as a guide, a runner can move out at full speed without wasting time checking map and compass.

As runners near the control, they begin searching for a catching feature crossing their path at 90 degrees to alert them to slow down and begin navigating more carefully. A catching feature can be any of the features described as potential handrails. Once runners locate the intersection of the handrail and the catching feature, they begin searching for the attack point: some relatively easy-to-find landmark close to the control. From there, runners follow a precise compass bearing for a distance measured off the map. To keep track of distance, they measure the length of their stride beforehand and then count paces until they locate the control.

Good route-finding involves more than identifying handrails, catching features and attack points. It also involves decisions about dealing with obstacles like hills, forests and brush: over, around or through?

As a rule of thumb, every foot of elevation gain takes as much time as running 12½ feet on the level. If your map's contour interval is 20 feet, then climbing one contour interval takes as much time as running 250 feet on the level. You can use this rule of thumb to estimate whether it's faster to go over a hill or around. Let's say going over the hill involves a climb of three contour intervals or the equivalent of 750 feet of horizontal travel, plus 250 feet of actual horizontal travel as measured on the map. The total is 1,000 feet. If going around the hill takes less than 1,000 feet of running, it's faster to go around. If it takes more than 1,000 feet, go over the top.

Another rule of thumb concerns the extra time required to run through vegetation and brush. Let's say it takes one unit of time for you to cover 100 yards on a good trail or road. You can then estimate that it will take you two units of time to cover the same distance through tall grass, three units of time through forest with light underbrush and four to six units of time through heavy underbrush.

If the idea of honing your route-finding skills by finding controls appeals to you, but you don't like the competitive aspect, you can attend nearly all meets and amble through the course (or a different one set up especially for people like you) at your own pace, locating the controls and checking them off your list. In orienteering jargon, you'll be known as a wayfarer or map-walker. Non-competitive orienteering is especially popular among families with small kids.

Appendix

Sources of Maps, Compasses, Altimeters and GPS Receivers

USGS maps are available from the USGS and many private map dealers. All private map dealers in a state are listed in that state's map index, available from the USGS. To mail-order maps, write to the USGS for a map index for the state you're interested in, plus the companion Catalog of Topographic and Other Published Maps for that state. Both are free. On the order form, list the map name, state and series or scale. Enclose a check, made payable to the USGS, for the amount due. Phone orders for less than 10 maps will be accepted, but orders by mail or fax are strongly preferred. Maps for the entire country, plus Antarctica, are available from:

USGS Information Services

P.O. Box 25286

Denver, CO 80225

Phone (303) 202-4700 or (800) HELP-MAP (435-7627)

Fax (303) 202-4693

The USGS has a web site where you can locate a private map dealer in your area or download an order form, which you can fax back. The address is:

http://www-nmd.usgs.gov/esic/to_order.html

To order Canadian maps, contact:

Canada Map Office

Phone (613) 952-7000 or (800) 465-6277

Fax (800) 661-6277

Ask for the name, address and phone number of the dealer nearest you, then purchase maps from that dealer either in person or by phone. You can also get contact information for the nearest dealer from the Canada Map Office's web site:

http://www.nrcan.gc.ca

National Geographic Maps publishes the Trails Illustrated series of topographic maps. These tough, waterproof maps, which are printed on plastic, cover many of the most popular hiking and backpacking areas in the western United States as well as 54 national parks and selected national forests nationwide. Trails Illustrated maps, which are based on USGS topos, are very convenient because they usually cover the entire area you'll traverse in a typical trip with one map. On a trip in Colorado's Indian Peaks Wilderness, for example, you can carry just one Trails Illustrated map instead of four or more 7½-minute USGS quads. The drawback is that the scale is smaller than on a USGS 7½-minute quad, so that the maps show less detail, which makes precision route-finding more difficult. If you can't find Trails Illustrated maps locally, contact:

Trails Illustrated

P.O. Box 4357

Evergreen, CO 80437-3746

Phone (303) 670-3457 or (800) 962-1643

Most map dealers and outdoor specialty shops carry compasses. If you have trouble finding what you want locally, you can order a compass or get the address of the nearest dealer from the manufacturer or distributor.

For Silva compasses, U.S. residents should contact:

Johnson Worldwide Associates

Outdoor Equipment Division—Silva Compasses

1326 Willow Road

Sturtevant, WI 53177

The automated dealer locator line is (888) 245-4986; to reach a human being, call (800) 572-8822.

Their web-site address is:
http://www.jwa.com

Canadian residents should contact:
JWA Canada, Inc.
4180 Harvester Road
Burlington, Ontario
Canada L7L 6B6
Phone (800) 263-6390

For Brunton compasses, U.S. residents should contact:
Brunton USA
620 East Monroe
Riverton, WY 82501
Phone (307) 856-6559
Their web-site address is:
http://www.brunton.com

Canadian residents should contact:
Brunton/Canada
6-637 The Queensway
Peterborough, Ontario
Canada K9J 7J6
Phone (705) 749-9327

For Suunto compasses, U.S. residents should contact:
Suunto USA
2151 Las Palmas Drive
Carlsbad, CA 92009
Phone (800) 543-9124

Canadian residents should use the same mailing address, but the following phone number:

(800) 776-7770

Thomen pocket altimeters are some of the best available. They're distributed by Suunto USA (see contact information above).

GPS receivers are becoming so widespread you can even find them at some mass merchandisers. If you can't find the model you want locally, contact the manufacturers.

For Eagle GPS units, contact:

Lowrance Electronics

P.O. Box 669

Catoosa, OK 74015

Phone (918) 437-6881 or (800) 324-1354

http://www.eaglegps.com

For Garmin GPS units, contact:

Garmin International

1200 East 151st Street

Olathe, KS 66062

Phone (913) 397-8200

http://www.garmin.com

For Magellan GPS units, contact:

Magellan Systems

960 Overland Court

San Dimas, CA 91773

To request literature via an automated system, call (800) 707-5221; to speak to a human being, call (800) 669-4477 or (909) 394-5000.

Their web-site address is:
http://www.magellangps.com

For Trimble GPS units, contact:
Trimble Navigation
2105 Donley Drive
Austin, TX 78758
Phone (800) 487-4662
http://www.trimble.com

For more information on the sport of orienteering, contact:
U.S. Orienteering Federation
P.O. Box 1444
Forest Park, GA 30051
(404) 363-2110
http://www.us.orienteering.org

Index